Fishing Boats Around Scotland

FISHING BOATS AROUND SCOTLAND

30 YEARS OF PHOTOGRAPHY

PETER DRUMMOND

Seiner/trawler *Sunrise* FR359 steaming into Peterhead.

SUNRISE
FR. 359

First published 2023

The History Press
97 St George's Place, Cheltenham,
Gloucestershire, GL50 3QB
www.thehistorypress.co.uk

British Library Cataloguing in Publication Data.
A catalogue record for this book is available from the British Library.

ISBN 978 1 80399 116 0

Typesetting and origination by The History Press
Printed and bound in Great Britain by TJ Books Limited, Padstow, Cornwall.

Trees for Life

Cover illustrations
Front: *Islander* BA316 heading for Ardrossan.
Back: *Unity* FR165 leaving Peterhead Bay.

Scallop dredger *Academus* BA817 in the River Dee at Kirkcudbright.

CONTENTS

INTRODUCTION

For over fifty years I've spent a big chunk of my spare time wandering around fishing harbours. For more than thirty of them, I've been armed with at least one camera. Two cameras are better than one as there are always occasions when black and white film produces a better result than colour. Here then is my selection of the best of my black and white pictures, augmented in places by pictures that were shot in colour but reproduce better in black and white. Some have been seen before: where this is so I think they're good enough to rate a second outing, though sometimes I've used a different shot out of a series of pictures to offer the reader something a little bit new.

I've set out this book in the form of a tour around some of the Scottish harbours, leaving Peterhead until last to have a big finish. That's where we boat photographers had our best times, with boats from all over the place running in and out, sometimes with remarkable frequency. Happy days now gone. On the other side of the country, it pleases me that I had the foresight to take quite a lot of pictures in south-west Scotland, which have added variety to this book, including illustrations of a number of former ring netters. I've tried to show vessels from as many different eras and of as many different types as possible.

While the main focus of the book is the pictures, I've included an appendix giving details of the boats illustrated. It's not comprehensive, but I hope it contains details that readers will find interesting. I've taken the builder to be the yard that handed the vessel over to the owner and not the sub-contractor who constructed the hull and basic superstructure, except where I have a particular reason for mentioning the involvement of multiple yards in the building process. The year of building is the year of delivery of a boat unless I have reason to think delivery was in the very early part of the year, which must necessarily place construction in the previous year. A boat's story stops when she leaves the United Kingdom or, as the case may be, leaves the UK for the last time, although a handful of boats have some bonus information included.

KIRKCUDBRIGHT

Kirkcudbright must be one of the few places where you can take a picture of a fishing boat very nearly in the town centre. Here, beam scalloper *Argonaut* BA858 sits obligingly for the camera.

Opposite page

Sophisticated 75ft scalloper *Kingfisher* BA810 heads down the River Dee. Her full shelterdeck is an unusual feature on a scalloper and she also has a mechanised catch-handling system.

Beam scalloper *Arcturus* BA862 heading upriver.

JK
KINGFISHER
BA810
BA810

BA 862
862 ARCTURUS

GIRVAN

This is *Osprey M* BA2; now a trawler, but in her heyday she was *Ave Maria* CY1, one of the renowned Eriskay ringers and among the last to fish the ring net commercially in the Minch in the winter of 1977–78.

Northern Irish visitor *Jubilee Star* B268.

Scalloper *Ellen Ann* BA359 outward bound after her annual repaint on Nobles's slip.

MAIDENS

The nearest I can manage to a picture of a ringer in the Maidens is this shot of *Brilliant Star* SH117, which at least looked a bit like a ringer and did spend some time at the Clyde herring fishing, albeit pair trawling with *Utopia* TT55. The picture was taken on 25 April 1998 when the boat had ceased fishing and was, I believe, awaiting conversion for non-fishing use.

AYR

Ocean Maid BA55 was built as an east coast seiner/trawler, came second-hand to Ayr to replace the former ring netter *Valhalla* BA165, and became one of Ayrshire's most successful prawn/white fish trawlers.

In her ring net days at the end of the 1960s and early ’70s, *New Dawn* BA18 was one of the biggest ringers, with an overall length a fraction under 65ft, and one of the most powerful with a 345hp Caterpillar engine. Her ring net and later herring pair trawling partner was *Fair Morn* BA295.

A feature of the Ayrshire fishing fleet in the 1990s was the arrival of a number of vessels of Danish design and/or construction, previously operated by expatriate Danish fishermen and usually based on Humberside. Sometimes the boats retained their English registrations when they moved to Scotland. This example is the Hull-registered *Falkenborg* H120.

Opposite page

One of the fleet of ringers once based at Maidens, *Ocean Gem* BA265 changed home port to Dunure in 1973. Her ring net partner at Maidens was *Pathfinder* BA252, while at Dunure she teamed with *Britannia* BA267.

When *Dawn Watch II* BA120 was new and named *Queen of the Isles* SY302, she belonged to Port of Ness, Isle of Lewis. She joined the Ayr fleet to trawl for prawns and white fish in the early 1970s.

Sea Harvester AR79 came to Ayr in 1981, part of a continuing movement towards bigger and more powerful boats by Ayrshire skippers. She was one of the regulars at the herring pair trawl in the later years of the Clyde herring fishery.

Spindrift BA220 was a 55ft seiner/trawler with a 270hp Scania Vabis engine, which gave her the power for herring pair trawling. Her partner at the herring was *Laurisa* BA145.

Another Danish-style anchor seiner that spent most of her working life fishing from Humberside and would later be adapted for trawling; this is *Binks* GY617.

Originally a trawler based at Kilkeel, Northern Ireland, and later converted for seining, the 65ft *Loyal Friend* INS450 was one of the last of a whole fleet of east coast seiners in the 60–70ft range that seasonally landed at Ayr.

Wanderer III INS161, once one of the local fleet at Ayr, was sold to east coast owners and became one of the last Moray Firth seiners to fish the west coast grounds.

A final look at visiting Moray Firth seiners. This is *Faithful* INS38.

Gratitude BA130 was always particularly welcome when she turned up for a picture due to her lack of any gutting shelter. She was built as a late 1960s ringer/trawler and that's exactly how she continued to look.

Former Mallaig ring netter *Westerlea* OB93, seen here as the prawn trawler *Huntress* BA93, was the last fishing boat to work from Ayr, continuing to land there even after the fish market moved to Troon in 1996.

The 40ft *Weston Bay* OB129 was one of a series of GRP (glass-reinforced plastic) stern trawlers built by Cygnus Marine Ltd, Penryn, Cornwall.

Saffron BA172 was built as a ringer/trawler at a cost of £63,000, but is best remembered for her time as part of a highly successful herring pair trawling team along with *Investors* BA117.

Girl May PD239 was once a Shetland drifter/seiner. Towards the end of her fishing career, she prawn trawled in the Clyde.

When *Silver Quest II* BA122 was *Wistaria* FR116 she was one of the most frequent east coast visitors to the Clyde herring fishing. When she made a permanent move to Ayr, she continued to spend as much time as possible pelagic trawling. Here she is rigged for the single-boat pelagic trawl.

SEA OTTER
BA 6

SHEARBILL
CY 571

Oruna BA20 was built to an entirely new design, with a large capacity and ability to work further afield than most other Clyde-based boats despite being only 60ft long. She fished as far away as Rockall.

Opposite page

Sea Otter BA6 came to Ayr second-hand from Brixham. She was used mainly for prawn trawling and scallop dredging, but also had a spell pair trawling for herring with *Frigate* BA138.

Shearbill GY571 has a distinctly Danish look to her hull, though she was actually built at Fraserburgh.

Fair Morn V BA19 was one of many boats around 55ft long built by yards in north-east Scotland for local owners that were then fitted out as trawlers or combination seiner/trawlers. Following her move to the west coast, she switched to prawn trawling.

When new, *Lunaria* AR82 was based at Bangor in Northern Ireland. Unusually, her engine room was forward. She was originally powered by a 430hp Baudouin engine and had a new 430hp Cummins installed in 1988.

Stormdrift BA187 was one of the last Clyde ringers active in the winter herring fishery in the Minch in 1976–77.

Opposite page

Spes Bona III BA107 was originally registered at Arbroath but based at Lowestoft. She was then sold to Arbroath owners, for whom she did well at the white fish trawl, and had a spell at Oban, before coming to the Clyde.

Northern Irish-registered visitor *Radiant Light* N243. When the picture was taken she was owned in Whitehaven. She will be best remembered as *Crystal Sea* N243, when she was a notable herring trawler.

BA.107
BA.107

N243
N243

Star of Peace II WK150 was another east coast seiner that switched to trawling and spen a while based in the Clyde, bu retained her Wick registration

Siskin AR302 was originally part of the fleet based at Thyborøn in Denmark. When she first came to the UK she was based at Grimsby, and had some success as a white fish pair trawler teamed with *Betty Ann* GY388.

British-registered but Spanish-controlled vessels are a controversial subject, but when a number of them began to use Ayr they gave me opportunities I otherwise would never have had to photograph boats that were once British side trawlers. *Breydon Mallard* LT131, once a Lowestoft sidewinder, is an example.

Flourish ME76 was another example of a vessel that spent her best days as an east coast seiner before switching to trawling in the later part of her career.

TROON

When new in 1964, the 59ft ringer *Britannia* BA267 was powered by a 220hp Caterpillar engine, which was one of the first Caterpillar engines to be installed in a new boat built in Scotland. It gave her a top speed of 11 knots, which made her about the fastest fishing boat based in the Clyde.

Opposite page

When *Andrew* BA698 was the *Olympic* UL16, she was one of the many seiners belonging to Moray Firth owners that fished from west coast ports, including occasional visits to Ayr. The boat's appearance was totally changed in 1988, the result of a major rebuilding by Coastal Marine Boatbuilders Ltd at Eyemouth that included a new wheelhouse, whaleback and gutting shelter.

Scottish-built, but Spanish-controlled, *Brisca* M16 arrives in port on a day of such dirty weather that only a black and white picture was possible.

BA 698

BRISGA
M 16
M 16

Spes Bona IV BA107 was originally built for owners in the Isle of Man. She has the look of a ring netter, though she never fished as one.

Silver Quest AR190 began as one of a number of east coast-based trawlers in the 50–60ft range built by Nobles at Fraserburgh. The design adapted well to prawn trawling on the west coast.

Constant Faith BA353 was part of the Lowestoft fleet before she moved to the east coast of Scotland, where she was based for a long time. She ended her career as part of the Ayrshire fleet.

Ocean Maid BA55 came to Troon second-hand from Northern Ireland and became the biggest vessel in the by then much-reduced Troon fleet. She replaced the older *Ocean Maid* BA55 (seen on p. 15).

AR.871

The 55ft *Trust* AR871 was built for Eyemouth owners as a seiner/trawler with a 270hp Volvo Penta engine.

Pathfinder OB181 was originally a Maidens ringer before becoming a Mallaig ringer. This picture shows her back in the Clyde engaged in prawn trawling, but still with her Oban registration.

When new, *Opportune* BA221 was *Cairnsmore* BA74, based at Garlieston on the Solway Firth.

Solstice II BF56 was one of a number of stern trawlers built by the Ailsa Shipyard at Troon and fitted out while berthed at the harbour's outermost jetty.

Strictly speaking, this isn't Troon, it's Barassie. At New Year 1999 when *Atlantic Challenge* PD197 was fitting out at the same berth where I photographed *Solstice II*, south-west Scotland was hit by extreme weather with wind speeds so high as to snap *Atlantic Challenge*'s mooring ropes and literally blow her out of Troon harbour. I saw her on 2 January 1999, aground on Barassie beach in bad, but less foul, weather, in which a photograph was just about possible. Fortunately, she was salvaged and completed. For the happy ending, see p. 109.

ARDROSSAN

Bon Ami BA104 is an example of the French style of small-to-medium steel stern trawler that has been increasingly seen in Scotland through second-hand acquisitions.

Opposite page

Spanish-controlled vessels also used Ardrossan as a base and there was always the hope of catching a picture of one that had once been a British side trawler. The 126ft *Menorca* AR777 falls into this category with some style, having been the highest-grossing trawler at Aberdeen in 1971 when she was *Japonica* A524.

The Campbeltown-based but Ballantrae-registered *Islander* BA316 heads in on a cloudy Ayrshire afternoon.

MENORCA
AR777

ARDRISHAIG

Tied up in the Crinan canal basin at Ardrishaig for the Christmas and New Year holidays and sitting quite nicely for her picture on 31 December 2001, *Shemaron* CN244 was the last Scottish boat ever to shoot a ring net (for scientific herring tagging in February and March 1990).

TARBERT

Outward-bound *Caledonia* TT34 once saved my day from being a complete failure; all attempts to take pictures of prawn trawlers leaving Campbeltown had proved futile as they left before there was enough daylight for photography. Tarbert on the way home was a different story.

Frigate Bird TT137 heading in to land.

Destiny TT279 also steaming in with her catch, albeit on the other side of the peninsula. She is heading for the pier in West Loch Tarbert.

CARRADALE

Few things could be more iconically Scottish than an ex-ringer heading into Carradale, once a ring net stronghold. *Village Belle IV* TT74 was built as a 60ft, 240-cran capacity ringer/trawler with a 240hp Kelvin engine.

CAMPBELTOWN

Little 40ft stern trawler *Adoration II* CN78 steams up the loch. Davaar Island looms large in the background.

Boy Andrew TT179 heading up Campbeltown Loch.

TAYINLOAN

St Ives-registered 40ft tangle netter *Louise Joanne* SS106.

MALLAIG

Vikingborg OB285 was a Danish-style anchor seiner, albeit one built in Buckie, and was based in Humberside before moving to Scotland to divide her time between trawling and scallop dredging.

CN.77
SPES NOVA

LK.70
LK.70
CONTEST

Mareather OB503 was due for her annual repaint as she headed into port, but she still came in at a good time (between the morning and afternoon Jacobite steam trains, which were my main targets that day) and gave me a very acceptable picture.

Opposite page

Spes Nova CN77 slipped for repainting. She was a fine example of a 40-footer built by Nobles of Girvan and a very powerful one indeed by the standards of her time (1972) with a 200hp Caterpillar engine. Not that any of what we all called 40-footers actually were 40ft long – they were all a few inches shorter to allow fishing within the 3-mile limit.

Contest LK70 has retained her original name and Lerwick registration after being sold away from Shetland, changing her home port to Mallaig in 2000 and to Troon in 2019.

When *Caralisa* OB956 was *Constant Hope II* KY100 she fished very well from Pittenweem. As *Caralisa*, she was a stalwart of the Mallaig fleet for years and was one of the local vessels that spent part of the winter pair trawling for sprats.

Hercules II LK438 was another fine example of a 40-footer built by Nobles of Girvan, though one from the generation before *Spes Nova* (p. 52), with a bit less power in the form of a 112hp Kelvin engine.

GAIRLOCHY

Silver Cloud II WK80 nears the locks at the south end of Loch Lochy, heading home to Gairloch via the Caledonian Canal after her annual repaint at Fraserburgh.

EYEMOUTH

Maryeared TT57 was originally part of the Tarbert ring net fleet and cost £18,000 to build. Her ring net partner was *Margarita* TT26, later replaced by *Catherine Ann* TT31. *Maryeared* and *Catherine Ann* continued to engage in pelagic fishing following the demise of the ring net and the pair trawled together for herring and sprats for as long as these fisheries were available to them.

Opposite page

Immanuel VII LH546 was, by British standards, an unusually large boat to have a hull constructed from GRP. She had a variety of home ports, having fished from the Isle of Skye, Peterhead, Orkney, Shetland, Northern Ireland, Eyemouth and Fleetwood.

I've rarely been in Eyemouth at the right time to catch boats on the move. One exception to this rule is French-style stern trawler *L'Ogien* K62 heading seaward. Local landmark Gunsgreen House is visible in the background.

THE HIP HOTEL
HIPPODROME
LH 546
Immanuel
LH 546

L'OGIEN
L'OGIEN
K·62
K·62

PITTENWEEM

St Adrian II KY202 was a highly advanced vessel for her time. A fraction less than 43ft long, she had the capacity and catching power of boats around 55ft long. She had two net drums, enabling her to carry two trawls to shoot as required. Nowadays we're fairly used to seeing quite small boats with two trawls wound around net drums, but in 1981 it was revolutionary.

GOURDON

The 41½ft *Ocean Vanguard* OB21 was originally based at Blyth, Northumberland. She came to Scotland in 1978 to be based in Anstruther, then later moved to Oban, and thereafter to Gourdon.

ABERDEEN

Ross Anne A155 leaves port against something of an urban background.

Opposite page

Alexanders A177, also outward bound.

Visiting English stern trawler *Armana* FD322 arrives on a misty day.

A.177
A.177
A.177

MARR
FD 322
ARMANA
FD 322

Radiant PD298 had to be photographed from further aft than I would have liked when outward bound, as the race out of Aberdeen to the photo location was one I didn't really win. Call it a draw.

One Monday a search for Kirkcaldy-registered seiners based at Aberdeen produced none other than *Argonaut IV* KY157, a definite candidate for the title of Scotland's champion fly-dragging seiner.

On the same Monday another redoubtable seiner from the north side of the Firth of Forth, *Arktos II* KY129, heads seaward.

talie B H1074, a Dutch beam wler turned beam scalloper, is ward bound. She didn't seem to hurrying into port, which was t as well as I first saw her ading in from north of the idge of Don and got into ce ahead of her.

Scalloper *St Amant* BA101 heads for port. She was originally one of a series of 59ft steel white fish trawlers that were based at Yorkshire ports.

Kirkcudbright-based *Academus* BA817 was purpose-built for scalloping. She has turned up at no fewer than five different ports where I was waiting with a camera. Well done *Academus*!

FRASERBURGH

Britannia BF178 spent over twenty years as a Lossiemouth seiner before being sold to Gardenstown owners in 1973. Her later years were spent trawling. She escaped destruction when decommissioned and was converted to a pleasure boat, but sadly flooded in Anstruther harbour in 2019 and broken up in 2020.

Serenity BF24 is a 68ft trawler. She had her original 440hp Kelvin engine replaced with a 440hp Mitsubishi in 1997.

Silver Fern FR416 had a thirty-six-year career as a trawler before undergoing a major conversion to dedicated scalloper by Macduff Shipyards Ltd, which included a new 478hp Caterpillar engine in 2018.

Our Hazel UL543 began as a Channel Islands crabber before joining the Scottish fleet.

I think we can tell from the water line of *Conquest* FR1 that her search for herring has been a successful one. Originally a purse seiner, her build cost of £70,000 was a huge figure by the standards of her time (1968). The picture shows her in her later years, when she was pair trawling with *Green Pastures* FR222.

Once a Dutch beam trawler, now an enormous beam scalloper, the 131ft-long *Isla S* DS1 heads out to sea.

Originally based at Westray, Orkney, and powered by a 930hp B&W Alpha engine, *Resilient* FR327 was one of several good-looking and very effective steel vessels built by James N. Miller & Sons Ltd at St Monance.

The 92ft beam scalloper *Verwachting* P845, registered at Portsmouth, but based at Kirkcudbright, makes an evening departure from Fraserburgh.

When *Kimara* FR176 was new she was one half of a particularly potent pelagic pair trawling team, the other half being sister vessel *Helena* FR178.

The only thing *Fladda-Maid* UL209 did wrong on a very fine morning was to arrive when the sun was still on the wrong side for an inward-bound shot. Definitely an occasion when I was glad to have a second camera loaded for black and white.

Fredwood II BA340 was the last wooden purpose-built scallop dredger to join the Scottish fleet before these vessels became almost exclusively steel.

Orion BF432 turned up shortly after *Fladda Maid*. The conditions were still the same and the black and white did the trick again.

The hefty *Kings Cross* FR380 was a fine example of a Norwegian-built pelagic trawler.

Forever Grateful FR249 was a purse seiner/pelagic trawler but, more unusually, was built in Spain as the shape of her bow tends to suggest.

Purpose-built scallop dredger *Georgia Dawn* INS140 leaves a busy Fraserburgh harbour in the evening.

Mia Jane W FR443 is an example of the French style of small stern trawler. She was built for French owners, worked from Eire for some time and then joined the Scottish fleet.

Provider II BF422 was one of several small steel stern trawlers built by skippers who often downsized from larger, older boats to more effective and economical designs.

Fertile FR599 belonged to Peterhead before being acquired by the Tait family of Fraserburgh to augment their family fleet of purse seiner/trawlers.

Kinnaird FR377 was an example of a type of vessel built in the 1990s to be capable of working west of Scotland in some exposed and deep waters, at a time when the indications were that the future of demersal fishing might lie in a switch from more traditional North Sea grounds.

Coronata II BF356 was one of the 73ft steel vessels nicknamed Sputniks that were series-built in the latter part of the 1950s and early '60s. Many, including *Coronata II*, had their working lives substantially extended by major rebuildings in the 1970s.

Modulus FR272 was one of a series of 86ft seiner/trawlers designed by Tynedraft Design Ltd of Newcastle upon Tyne, almost all of which went through difficult and greatly extended building processes due to the economic crisis in the 1970s (see Appendix). This vessel's fortunes changed completely when she was sold to Fraserburgh owners in 1978 and had RSW (refrigerated sea water) tanks for pelagic fishing installed in 1979. She was one of Scotland's leading pair trawlers and her longest-serving partner was *Uberous* FR50, which was built as the purse seiner *Comrade* FR122.

When *Suilven IV* BCK366 was new and named *Chelaris* BF16, she was a highly successful white fish pair trawler based at Macduff that was teamed with *Tranquility* BF7.

Daystar BF151 was one of many aft-wheelhouse, twin-rig steel trawlers that were built by Macduff Shipyards Ltd, the later corporate incarnation of Macduff Boatbuilding & Engineering Co. These vessels became mainstays of the east coast prawn trawling fleet.

The hull and basic superstructure that eventually became *Ocean Harvest* BF145 went through the same traumatic building process as sister vessel *Modulus* FR272 (p. 75). Like her sister, she became part of the Scottish pair trawling fleet that alternated between pelagic and white fish operations, with her regular partner *Bountiful* BF79.

Zara Annabel BCK126 began as a Dutch beam trawler. She was acquired by English owners for continued operation as a beamer, but in her later days on the Scottish register, she has been used for scallop dredging.

A formidable visitor from the Peterhead fleet outward bound to start her North Sea herring season. This is the 215ft pelagic trawler *Quantus* PD379.

Pindarus FR159 was another vessel built to the Tynedraft 86 design, one of the variant with a 600hp B&W Alpha engine. As *Unity* PD209 she was one of the vessels that alternated between white fish and pelagic operations and this continued after she became *Constant Hope* PD209 and later *Paramount* PD209.

This book has no conscious bias in favour of boats built at Macduff, but the port has been the birthplace of so many of them that sheer weight of numbers boosts their chances of making the list of best pictures. *Bracoden* BF37 is one of the steel-hulled examples.

Davanlin FR890 is another vessel built in Macduff (this time of wood) that has given sterling service, including brightening up what was proving to be a rather dull day for me before this picture was taken.

Twin-rig prawn trawler *Replenish* BF28. Subsequent to this picture being taken, she was lengthened by 15ft by MMS Ship Repair & Dry Dock Ltd in Hull in 2008.

The 87ft seiner/trawler *Carvida* FR457 was another fine-looking steel vessel built at St Monance. She had an 860hp Mirrlees Blackstone engine.

Winner FR265 was originally a Shetland drifter/seiner. She is seen in her later days, based at Fraserburgh and rigged for trawling.

Aquinis BA500 was one of the nomadic scallop dredgers owned at Kirkcudbright that turned up at ports all around the British coast and often made a welcome addition to the day's photographs.

SUNBEAM
SUNBEAM
FR 487
FR 487

GRATEFUL
FR 249
GRATEFUL
FRASERBURGH
FR 249

MACDUFF

Locally owned top-earner *Audacious* BF83 heads into Macduff.

Opposite page

Sunbeam FR487 is less obviously Spanish built than *Forever Grateful* FR249 (p. 72), although both ships came from the same builder. *Sunbeam* is seen after being lengthened by 26ft by Baatbygg AS, Måløy, Norway, in 2004, an operation that significantly altered her forward profile as well as extending her amidships.

Let's have a big finish at Fraserburgh: this is the 230ft pelagic trawler *Grateful* FR249.

West coast visitor *Intrepid* BA100, in Macduff for an overhaul and repaint, comes into the port in monsoon-like conditions.

BUCKIE

Nordic Prince BCK18 was a most unusual vessel by Scottish standards. She was a 108ft auto-liner with catch-freezing facilities powered by a 600hp B&W Alpha engine. She came second-hand from Norway in 2000.

LOSSIEMOUTH

How I wish I'd for once focused less on the big new boats in the Scottish fleet and snapped some of the older vessels on the move, like the seiners at Lossiemouth. However, long after the demise of the locally based seiners, *Clarness* INS51 gave me a picture of a trawler making her way out of the port.

BURGHEAD

I also found it hard to catch a boat on the move at Burghead, but *Alex Watt* INS163 chose the right moment to go to sea during one of my relatively rare visits.

PETERHEAD

Accord PD90 was originally the Shetland purse seiner *Altaire* LK429. For around ten years, starting in 1994, she was always running in and out of Peterhead during the North Sea herring season, when she was usually pair trawling with *Unity* FR165 (p. 157).

Opposite page

Della Strada CY158 began her career as an 87ft purse seiner. The picture shows her in her later career as a pelagic pair trawler, by which time her appearance bore little resemblance to how she looked when new, the result of lengthening by 14½ft in 1977 and a major rebuilding, including lengthening by 23ft, in 1984.

Anyone who looks at Scarborough visitor *Allegiance* SH90 and thinks the boat has been lengthened is absolutely correct as Parkol Marine Engineering Ltd at Whitby put a 19ft section into her midships and also extended her stern by 4¾ft in 2011.

CY.158

SH.90
ALLEGIANCE
SH.90

Danish-style and Grimsby-based visitor *Tino* GY203 arrives on a typically misty morning. This vessel did particularly well as one of the Humber port's fleet of white fish pair trawlers, usually teamed with *Samantha* GY327.

When the 80ft seiner/trawler *Favonius* PD17 was delivered in 1969, she was the first of what was to become a procession of big steel vessels built for the Scottish inshore fleet.

Starlight PD150 was another Tynedraft 86-footer. She regularly teamed up with sister *Constant Friend* PD83, pair trawling for herring, mackerel and later white fish. A third Tynedraft 86-footer, *Unity* PD209, was an additional partner at the herring.

One of the last boats to join Peterhead's once sizable fleet of boats in the 40ft class and the last survivor of the 40-footers was the seiner/trawler *Serene* PD58.

BARBARELLA
PD.134

FAITHFUL
FR.129
FAITHFUL
FR.129

When *Evening Star* PD1022 was built as *Tobrach N* TN2, she was a highly innovative vessel, designed around an automated catch-handling system with a full-length shelterdeck and her main engine mounted forward.

Opposite page

Barbarella PD134 was originally owned in Lyme Regis and moved to Peterhead during the boom years that were the first half of the 1970s. Some of the second-hand acquisitions for the Scottish fleet – like the 48ft-long, steel-hulled *Barbarella* – were non-traditional types, to put it mildly.

Faithful FR129 had an extremely successful pair seining partnership with *Crystal River* FR178 that lasted from 2004 to 2021, when both boats were replaced with brand-new sister vessels.

This is an excellent example of the wooden stern trawlers built in the late 1960s. *Cardanel* BCK9 was 65ft long with a 230hp Gardner engine.

This picture of *Silvery Sea* OB245 shows a much-modified purse seiner that started out as an 87ft boat. These rebuilding jobs tended to be done by Dutch or Scandinavian yards but *Silvery Sea* was modified by Richards (Shipbuilders) Ltd at Great Yarmouth, who lengthened the vessel by 35ft, raised her wheelhouse and fitted her with a shelterdeck in 1985.

Sunbeam FR478 originally belonged to Swedish owners. Her original name was *Abba*, presumably because her owners were Abba AB, something that nearly caused a difficulty for a group of extremely well-known Swedish musicians who needed the consent of the company to use the same name.

Iolanthe PD131 was once *Replenish* BF84, part of one of Scotland's most successful herring pair trawling teams along with sister vessels *Bracoden* BF37 (*not* the *Bracoden* shown on p. 79) and *Steadfast Hope* FR43.

The pair seiner *Lapwing* PD972 arrives on a very good day indeed. If only we got more days like this one!

PD.972
LAPWING
B
28

The highly successful *Ajax* INS168 was the first vessel built to Campbeltown Shipyard's 85ft design.

Starina INS801 was a 75ft seiner/trawler powered by a 500hp Kelvin engine. She was re-engined with a 650hp Kelvin in 1986.

Advance II INS77 was one of Campbeltown Shipyard's extremely successful 87ft design. She had a particularly productive pair seining partnership with sister vessel *Andromeda II* INS177.

Between 1992 and 1994, a number of sizable steel trawlers were bought second-hand from Swedish owners by crews in the north-east of Scotland and in Shetland. The 90ft *Sette Mari* LK257 was one of the vessels that moved to Shetland, arriving in 1992.

An example of a large wooden boat built in Eire, *Vertrauen* BF450 was originally *Girl Stephanie* G190, an 87ft trawler with an 850hp Caterpillar engine.

When *Provider* PD250 was new, she was *Rose of Sharon III* LH56, part of a fleet of vessels owned by the Moodie family from Port Seton, who were among the top fishermen from south-east Scotland.

Ocean Bounty PD182 has split winches mounted in her bows, enabling her to haul trawls forward through a tunnel beneath her wheelhouse and creating much-increased space for gear handling and repairing.

When *Fairwind* FR317 was new and named *Wave Crest* LK276, she was an innovative vessel designed for purse seining and readily adaptable for white fish trawling or seine netting between herring seasons. She was lengthened by 15ft by Baatbygg AS, Måløy, Norway, in 1978 and the shelterdeck was fitted by Malakoff Ltd in Shetland in 1983.

Alert FR396 was built as the Shetland purse seiner *Charisma* LK362 in 1979. She was lengthened by a hefty 49ft by Longva Mek Verksted, Haugsbygda, Gurskøya, Norway, in 1985.

Morning Dawn PD359 was a fairly uncommon instance of an investment by Scottish fishermen in a former Dutch beam trawler to carry on beam trawling for prime fish species. We've had no lack of second-hand Dutch beam trawlers in Scotland, but usually they are either converted for otter trawling or used as scallop dredgers.

Fruitful Harvest III PD247 was built as one of the smaller seiners for the Peterhead fleet. Emulating the days when a disproportionate number of seine netters were powered by Gardner engines, she had a 230hp Gardner that made her very economical for the seine net and she lasted at the fishery from Peterhead until 2007.

Jasper III PD174 was originally *Pisces* A193, one of the very successful class of 86ft side trawlers built for Aberdeen owners from the late 1960s through much of the '70s. She served the side trawling fleet from 1971 to early 1985, when she was fitted with a three-quarter-length shelter and converted to fish as a seiner, light-bottom trawler and pair trawler.

One of Herdies' wee stern trawlers – translation: one of the 56ft steel stern trawlers built by Herd & Mackenzie at Buckie. *Silver Star* BCK131 was one of the earlier batch with no whaleback even after the later fitting of a shelterdeck.

Wave Crest BCK217 was another of the steel stern trawlers built by Herd & Mackenzie and here she positions herself nicely to give a different angle to the picture of *Silver Star*. *Wave Crest* was one of the later variations of this type of vessel, with a whaleback from new, though the shelterdeck was added later.

In the early 1990s, Campbeltown Shipyard delivered three of its 87ft class, fitted out as demersal trawlers, to Orkney skippers. *Arkh Angell* K616 was one of these. On this occasion she obliged me by heading into Peterhead for her annual repaint when I was lurking on the south breakwater.

Convallaria VI BF58 was one of the summer regulars that ran in and out of Peterhead during the North Sea herring season, providing numerous photographic opportunities.

For a while Peterhead was a regular landing place for a handful of French stern trawlers that fished as white fish pair trawlers. This member of the French connection is *Cap Saint Georges* BL924675.

Opposite page

Diligent PD314 was built as a purse seiner (*Coronella* BF277) and later lengthened by around 20ft. Eventually she became too small to work effectively as a purser, so she changed from being, by the standards of the time, a small purser to a big pair trawler. The move was a good one because she formed one half of an ace pair team with *Sparkling Star* PD137 (seen below), fishing for herring and mackerel during pelagic seasons and white fish outside of them.

Sparkling Star PD137 was regarded by many, me included, as the best looking of the pair trawlers built in the mid-1970s. She began life as an 87ft seiner/trawler with a 750hp Mirrlees Blackstone engine. Her appearance was transformed by her builders, K Hakvoort Scheepsbouw NV, Monnickendam, Netherlands, in 1979 when she was lengthened by 20ft and fitted with a shelterdeck and three RSW tanks with capacity for 140 tonnes.

DILIGENT
PD314
PD314

PD137
SPARKLING STAR
PD137

Resplendent PD225 was a large stern trawler of a type completely alien to Scottish inshore fishermen, but intended to be the ultimate solution to the problems of working in the deep and totally exposed waters west of Scotland.

Boy Ken II DO5 is one for the ring netter enthusiasts, though you need to take a moment to think why – the new wheelhouse fitted by the Booth W. Kelly Ltd shipyard at Ramsey in the Isle of Man in 1994 probably doesn't help! This was once the Campbeltown ringer *Maureen* CN185, dating all the way back to 1949.

Atlantic Challenge PD197, seen following her salvage, repair and delivery, in much happier circumstances than in the photo on p. 41.

Purse seiner *Krossfjord* BF70 makes an evening departure. She was a second-hand addition to the Scottish purse seine fleet, arriving from Norway in 1984.

Taits FR228 was an extremely successful purse seiner/trawler. Her ability to catch herring is readily apparent from her position in the water.

Pelagic trawler *Resolute* BF50 went through a very useful phase of constantly turning up at Peterhead when I was hanging around the harbour with a camera. As she seemed to be offering to have her picture taken, I wasn't going to turn her down.

Eilean Croine CY190 was an example of a stern trawler that originally operated from Eire. She was sold to owners in Castlebay on the Island of Barra in 1988. As built, the vessel was 86ft long, but she was lengthened by a hefty 30ft by Killybegs Shipyard Ltd in 1983.

Shemara PD314, originally *Brendelen* SO709, began as an 88ft stern trawler powered by a 950hp Caterpillar engine. The picture shows the boat that existed after being lengthened by 38ft by Scheepswerf Visser BV, Den Helder, Netherlands, in 1986 and another major refit including a new 1,500hp Caterpillar engine by Karstensens Skibsvaerft AS, Skagen, Denmark, in 1991.

Pathway PD165 by twilight. Peterhead has had five purse seiner/trawlers named *Pathway*, one registered PD65 and the other four PD165. Illustrated here is the 189ft-long version built in 1998, catching the last of the sunlight in the bay.

Pathway PD165 in the afternoon. This is the biggest *Pathway* – so far – a 258ft purse seiner/trawler powered by 7,000hp Wärtsilä engine.

Visiting Norwegian purse seiner *Radek* H-15-AV heads seaward on a misty evening.

Credit for this picture is due solely to Grimsby visitor *Jubilee Spirit* GY25. It can't be easy to time an arrival in port just as the sun finds a gap in thick fog to break through.

Lomur LK801 was 65ft long when she was built in Norway for Icelandic owners. Lengthened by 20ft in 1991, she was bought second-hand by an Orkney partnership at the end of 1994. She was equipped to freeze catches at sea and her Orcadian owners used her to catch and freeze prawns.

A Dutch-style and UK-flagged pelagic pair trawler with freezing capacity, *Wiron 5* PH1100 heads in with mackerel.

Harvester PD98, originally *Devorgilla* BA67, was just below 70ft in overall length, an unusually large vessel to be built by Nobles at Girvan. She was first based at Stranraer, had a spell at Eyemouth and then moved to Peterhead, where she had a very successful white fish pair trawling partnership with *Ocean Harvest* PD198, ex *Sparkling Wave* B240.

Ocean Harvest PD198 demonstrates my point about the team referred to on the opposite page fishing well. This is a different *Ocean Harvest* PD198, an 85ft boat purchased by owners who were investing the results of their success in their industry.

Another instance of multiple boats having the same name and number. Peterhead has had four purse seiners called *Vigilant*: one registered PD452, one PD165 and two PD365. This *Vigilant* PD365 was a 133ft vessel built in 1980.

Opposite page

Meanwhile, this *Vigilant* PD365 was the largest of them at 165ft. She was built in 1995.

Glenugie PD347 was built as the 79ft wooden seiner *Prestige* INS305 for Moray Firth owners. When I took this picture she was seining for Peterhead owners and doing well at it too.

VIGILANT
PD-365
PD-365

PD.347
PD.347
PD.347

Magnificent PD250. The 100ft vessel was a former Lowestoft drifter/trawler that was built as the *Ethel Mary* LT337, sold to Fraserburgh owners in 1969 and became *Golden Promise* FR186 in 1970. Following an unsuccessful fishing venture in South Africa that began in April 1970, the vessel returned to Scotland in the late summer of 1972 and had a big refit to enable her to work as a seiner/trawler.

Opposite page

Sea Lady TN20 is a former Dutch beam trawler now scalloping. This vessel is 107ft long and was originally powered by an 800hp Mitsubishi engine until a replacement 1,000hp Mitsubishi was installed in 2002.

Leanne PD345, originally the Northern Irish-owned *Benaiah II* B350, was 70ft long with a 500hp Kelvin engine until James N. Miller & Sons Ltd at St Monance went to work on her over the winter of 1988–89. She was lengthened by 5½ft and fitted with a new 535hp Kelvin engine.

TN-20
TN20

PD.345
PD.345
LEANNE

An example of a sizable wooden vessel built in Northern Ireland, *Scotia* PD233 was built in Bangor, Co. Down, for Northern Irish owners and came second-hand to Peterhead.

Sustain PD378 was one of the first batch of 86ft steel vessels built to the Tynedraft 86 design to be delivered. She was then *Morning Dawn* PD195 and powered by a 637hp Mirrlees Blackstone engine.

Altaire LK429 heads seawards; she is a 250ft pelagic trawler powered by a 10,728hp Wärtsilä engine.

Aalskere K373 was built to be capable of working in deep water, but at 111ft overall, she was a step up from other vessels, with a registered length below 80ft and an overall length between 85 and 90ft.

Fear Not II PD354, originally *Sea Spray II* PD245, was one of the smaller units of the Peterhead fleet at 60ft overall, although her builders managed to fit quite a lot of boat into that length. When white fish quotas became tight and prawn trawling in the North Sea became a more attractive proposition, many skippers were attracted to high-capacity smaller boats that could take part in the prawn fishery more economically.

D.354
Fear Not.

Courage BF212 was originally the Shetland purse seiner/trawler *Azalea* LK396. It's hard to imagine that her original overall length was 128ft; Karstensens Skibsvaerft AS, Skagen, Denmark, lengthened her by a massive 73ft in 1987.

Opposite page

Prowess BF720 was built as *Paula* SO720, one of a fleet of five purse seiner/trawlers in the 130ft range built for owners in Eire at the start of the 1980s. Three, including *Prowess*, came from Dutch builders, while the other two came from Norwegian yards. Winter weather conditions west of Ireland proved too wild for purse seining, but the vessels did spectacularly well as pair trawlers. She was lengthened by 39ft in 1985.

Moremma PD135 was built for owners at Buckie, where she was based for ten years. She has now been Peterhead registered for over twenty years.

BF720
PROWESS

PD.135

Sunrise FR359 is one of the later generation of 80ft boats built by Campbeltown Shipyard. Delivered in 1984, she has kept the same name and number for her entire life.

Shalimar PD303 was built as *Salamis* PD142, one of the earlier examples of the seiner/trawler variant of the Spinningdale class of trawler. She was delivered in time to take part in the herring fishing, albeit in the declining years of these fisheries, which either closed completely or became tightly restricted in the late 1970s. Her partner was sister vessel *Budding Rose* PD84.

Elysian PD444 dates back to 1960, which means I was especially pleased to find this picture of her, again regretting my tendency to pay little attention to older types when there were a great deal of fishing vessels around.

Renown FR246 is an 80ft seiner/trawler. When new she had a 671hp Deutz engine, but this was later replaced by a 670hp Caterpillar.

WESTWARD
BF.350
WESTWARD
BF.350

PD.118
PD.118

Beam trawler/scallop dredger *Vandijck* BM362 was a long way from her base at Brixham in Devon when this picture added some welcome variety to the morning's activities.

Opposite page

Westward BF350 was originally *Sheanne* SO716, a Norwegian-built example of the Irish super trawlers (p. 127). As illustrated, she was a good bit bigger than her original 133ft, having been lengthened by 42ft by Eide Contracting AS, Høylandsbygd, Norway, in 1985.

Nordfjordr PD118 was originally built for Orkney owners at a time when the Orcadian fleet was extensively updated by the addition of several large, powerful and very effective white fish catchers.

Success PD789 was originally the *Gleaner* CN284, a 40ft seiner/trawler with a 110hp Gardner engine built by Thames Launch Works Ltd, Twickenham, in 1967. This was the first in a line of small steel boats built for Scottish owners, initially by Thames Launch Works and then by Campbeltown Shipyard Ltd, which grew into one of Scotland's leading builders of steel fishing vessels.

Opposite page

A visitor from the south-east of Scotland. This is 70ft Eyemouth-based pair trawler *Bonaventure* LH111.

And here's the other half of the team, *Bonaventure*'s sister vessel *Heatherbelle VI* LH272.

LH.111
LH.111

LH272
LH272

Another visitor from the south side of the Firth of Forth. Port Seton-based *Sharon's Rose* LH317.

Danish-style *Viking Warrior* PD188 was originally built for owners in Folkestone and was another non-traditional addition to the Peterhead fleet in the 1970s.

Ocean Way PD465 was unusual for a Scottish purse seiner/trawler because she had freezing capacity in addition to being able to carry catches in RSW tanks.

When *Scottish Maid* B417 was registered BF317 she was the first wooden boat built in Scotland to be fitted with RSW tanks for storing catches of herring and mackerel from the beginning. She was also the first with a Hedemora engine – an 840hp unit.

OPPORTUNUS
PD.96
PD.96
PD.96

KINGS CROSS PD 365
KINGS CROSS
PD 365
KINGS CROSS
PETERHEAD

Ocean Star FR894. You are seeing the result of three major rebuildings. Karmøy Mek Verksted, Kopervik, Norway, lengthened the vessel by 20ft and fitted three CSW (chilled sea water) tanks in 1974; Kopervik Slip AS, Kopervik, Norway, installed three more fish tanks, an RSW system and shelterdeck in 1979; finally, she was lengthened by 24ft by Maaskant Machinefabrieken, Bruinisse, Netherlands, in 1988.

Opposite page

Opportunus PD96, built as *Ajax* INS168, was an example of a late 1960s Moray Firth seiner/trawler. While we tend to remember her as a seiner, she spent the winter of 1970–71 pair trawling for herring in the Minch as part of a five-boat team of top-earning seine netters, the other four being *Horizon* INS21, *Forthright* KY173, *Steadfast* KY170 and *Lothian Rose* LH38.

Inward bound during the North Sea herring season is the 258ft purse seiner/trawler *Kings Cross* PD365. She had a 7,000hp Wärtsilä engine.

Fair Morn PD224. Thomas Summers & Co. at Fraserburgh built a number of small fishing boats referred to as yoles or yawls, or just plain bonny wee boats. *Fair Morn* PD224 was 33ft long and powered by a 42hp Gardner engine.

Opposite page

Christina S FR224 inward bound. She was a 170ft purse seiner that came to Scotland second-hand from Norway in 1987.

Christina S FR224 inward bound and a little further into the bay, but not the same boat. This one was a 189ft pelagic trawler built to replace the *Christina S* seen above.

FR 224
FR 224
CHRISTINA S

CHRISTINA S
FR 224
FR 224 CHRISTINA S

From 1994 onward, Spanish shipyards began delivering steel vessels to Scottish owners, often hefty vessels designed to be capable of fishing in deep, exposed western waters to take pressure off some of the more traditional Scottish fisheries in the North Sea. *Fair Morn* INS308 was one of these. She was a 91ft trawler with a 990hp Caterpillar engine.

At a time when foreign yards were delivering chunky steel boats for trawling, usually at very competitive prices, it was still possible to find British-built examples of the general type. One of these was English-built *Venture* PD72.

White fish pair trawler *Ryanwood* FR307 was built as *Faithful* UL179 in 1988, when she was the first steel vessel delivered by Jones Buckie Shipyard Ltd. Vessels built at this time showed a move away from wood hulls to steel, and to much heftier vessels than the preceding generation of seiner/trawlers. They were like an intermediate stage between the traditional boats and the chunky designs of the 1990s.

Regina Caeli CY58 was built as a purse seiner, but was later extensively modified and ultimately converted to a large pelagic trawler. When built as *Convallaria V* BF58, she was 88ft long. She was lengthened by 20ft in 1977 and 16ft in 1987, both times by Maaskant Machinefabrieken, Stellendam, Netherlands, who also installed a new 1,800hp Stork Werkspoor engine the second time. The conversion for trawling was done at Buckie in 1998.

The 233ft-long *Chris Andra* FR228 is one of Scotland's twenty-first-century pelagic trawlers and one of my regulars when I'm looking for pictures of these ships during the North Sea herring season.

Boy John INS110 had a fine record for catching fish and, in my experience, an equally good one when it came to turning up at good times for pictures, e.g. that tricky feat of finding a patch of light in the fog.

Ventura INS334 was a fine example of the big wooden boats produced by Richard Irvin & Sons Ltd at Peterhead in the 1970s. Originally the *Graceful* PD133, she was an 80ft seiner/trawler with a 495hp Mirrlees Blackstone engine.

Paragon V PD786 was an example of a 1990s high-capacity steel vessel and was the last boat to come from Jones Buckie Shipyard Ltd, which had been incorporated as long ago as 4 January 1916. Sadly, the yard was unable to deliver the completed vessel as it succumbed to the decline of British shipbuilding, though the business was combined with that of the neighbouring Herd & Mackenzie yard to form Buckie Shipyard Ltd. Meanwhile, *Paragon V* would be completed by Dunston (Ship Repairers) Ltd of Hull.

PD.981

Ability PD981 and visiting cruise ship *Deutschland* provide the kind of contrasting vessels shot for which Peterhead was always particularly good.

VOYAGER
N905
VOYAGER
N905

INS.185
INS.185

Falcon H119 was launched as *Inter Nos II* KY57 and was the first of a new, extremely effective and very good-looking class of large steel seiner/trawler built by James N. Miller & Sons Ltd, St Monance.

Opposite page

A very impressive visitor from Northern Ireland – the 231ft pelagic trawler *Voyager* N905.

Strathnairn INS185 was originally *Janet Sate* FR107, one of the largest white fish pair trawlers to operate from Grimsby. She was extremely successful when new, but was working at a time when the Grimsby pair trawlers had become as good at catching cod as it was possible for them to be, and quota restrictions on catches began to bite. These eventually led to her sale to Scottish owners.

Fisher Rose FR896 was an 80ft vessel from Campbeltown Shipyard that fished very well as a seiner. The buildings in the background from centre to left form part of Peterhead's largest guest house, formally known as HM Prison Peterhead.

Opposite page

Peterhead is regularly visited by a couple of very good-looking and well-kept vessels from Whitby. Here comes *Our Lass III* WY261.

And here comes sister and pair seining partner *Victory Rose* WY34.

OUR LASS III
WY·261
WY·261

VICTORY ROSE
WY 34
WY 34

Serene FR491 had a very successful career as a Shetland purse seiner between 1978 and 1987, firstly as *Antares* LK491 and then as *Serene* LK491. Her effectiveness as a catcher was increased when she was lengthened by 22ft by Karmøy Mek Verksted, Kopervik, Norway, in 1981. The vessel continued to fish as a purse seiner from Fraserburgh until 1997, when she was adapted to work as a pelagic pair trawler.

Even in a rather work-worn condition, *Regent Bird* BCK110 is still a good example of the big wooden boats built by Richard Irvin & Sons Ltd, Peterhead, in the early 1960s. Originally *Graceful* PD343, she was 79ft long and powered by a 200hp Gardner engine. She changed her name when a new *Graceful* was built by the same yard in 1974 (p. 143). She was fitted with a new wheelhouse and a new 415hp Kelvin engine by Macduff Boatbuilding & Engineering Co. in 1979.

Whitby visitor *Sophie Louise* WY168 gives the impression of having been lengthened. Coastal Marine Boatbuilders Ltd, Eyemouth, extended her by 13½ft in 1998.

Another representative of the fishing industry in the Lothians. The aptly named *Lothian Rose* LH380 was an 80ft seiner/trawler with a 575hp Callesen engine.

FR77
FR77
CHALLENGE

K.391
K.391
RIVO 1

Acorn INS237 was the first of a series of large steel seine netters built in Denmark for Scottish owners.

Opposite page

Challenge FR77 was built as the 128ft purse seiner/trawler *Zephyr* LK394. The vessel became one of the largest units in the pelagic fleet after being lengthened by 55ft by Larsnes Mek Verksted, Larsnes, Norway, in 1985.

Orcadian visitor *Rivo I* K391 is seen after being lengthened by 10ft by Wear Dockyard Ltd, Sunderland, in 1989.

Fruitful Bough PD109 was one of the vessels featured in the BBC television series *Trawlermen*. Here she takes a break from her screen career to have her picture taken by a less-prestigious cameraman, i.e. me.

Opposite page

Lunar Bow PD265. Peterhead has had no fewer than seven purser/trawlers named *Lunar Bow*, and six out of the seven have been PD265, which certainly makes it easy to get them mixed up. This is the 166ft 1996 version heading into the bay.

The 200ft 2000-built *Lunar Bow* PD265 is also inward bound. To add nicely to the scope for confusion, both this vessel and her predecessor were sold to Icelandic owners and both became *Asgrimur Halldorsson* SF250.

LUNAR BOW
PD-265
PD-265
LUNAR BOW

LUNAR BOW
LUNAR BOW
PD-265
PD-265
ET

ADENIA LK-193
LK-193 ADENIA

CELESTIAL DAWN
BF 109
BF 109

Unity FR165 had an extremely successful purse seining career as *Pathway* PD165. She also did well as *Unity*, including a long-running and very effective pelagic pair trawling partnership with *Accord* PD90 (p. 88).

Opposite page

A most impressive visitor from Shetland, the 229ft pelagic trawler *Adenia* LK193. She has a 7,240hp Bergen engine.

Celestial Dawn BF109 is an obliging vessel that always seems to turn up to have her picture taken when I'm at Peterhead or Fraserburgh. She has more than earned her place in this book!

Built in 2017, the 90ft-long *Faithlie* FR220 is one of the new generation of forward wheelhouse seiners.

APPENDIX: LIST OF VESSELS ILLUSTRATED

KIRKCUDBRIGHT

Argonaut BA858, built by Welgelegen Scheepswerf & Machinefabriek, Zoutkamp, Netherlands, as *Reseda* HA106, 1992. Became *Argonaut* BA858, 2013.

Kingfisher BA810, built by Hepworth Shipyard Ltd, Paull, Hull, 1998.

Arcturus BA862, built by Maaskant Machinefabrieken, Stellendam, Netherlands, 1993. Note: vessel never delivered to originally intended owner as they were unable to complete building contract. Lay incomplete for lengthy period until acquired by UK owner and fit out completed by contractors in UK.

GIRVAN

Osprey M BA2, built by Weatherhead & Blackie, Dunbar, as *Ave Maria* CY1, 1968. Became *Osprey M* BA2, 1983.

Jubilee Star B268, built by Bangor Shipyard Ltd, Bangor, Co. Down, Northern Ireland, 1976. Capsized and sank near Dunoon, February 2009.

Ellen Ann BA359, built by Miller Methil Ltd, Methil, as *Ocean Star* FR894, 2003. Became *Sparkling Waters* FR894, 2011; *Charlotte Ann* OB222, 2014; *Ellen Ann* BA359, 2016.

MAIDENS

Brilliant Star SH117, built by J. & G. Forbes & Co., Sandhaven, 1950. Converted for non-fishing use 1998.

AYR

Ocean Maid BA55, built by James N. Miller & Sons Ltd, St Monance, as *Ocean Triumph* KY56, 1973. Became *Ocean Maid* BA55, 1978. Decommissioned 2019.

New Dawn BA18, built by Herd & Mackenzie, Buckie, 1968. Decommissioned 2002.

Falkenborg H120, built by Glyngøre Skibs-og Baadebyggeri, Glyngøre, Denmark, 1971. Total loss after running aground on Black Rocks in Sound of Islay, 29 June 1994.

Ocean Gem BA265, built by Alexander Noble & Sons Ltd, Girvan, 1964. Ceased fishing 2006 and sold for conversion to pleasure boat.

Dawn Watch II BA120, built by John Watt & Sons, Banff, as *Queen of the Isles* SY302, 1961. Became *Queen of the Isles* BA120, 1971; *Harmony* BA120, 1976; *Dawn Watch II* BA120, 1984.

Sea Harvester AR79, built by Jones Buckie Shipyard Ltd, Buckie, as *Sea Harvester* N239, 1968. Became *Sea Harvester* AR79, 1982; *Sea Harvester* N822, 1995.

Spindrift BA220, built by Alexander Noble & Sons Ltd, Girvan, 1974. Became *Sharon Rose* SY190, 1999; *Boy Darren* CY190, 2009; *Spindrift* BA220, 2016. Constructive total loss following wheelhouse/galley fire in Campbeltown harbour, February 2019. Lying derelict in Girvan as of 2022.

Binks GY617, built by Herd & Mackenzie, Peterhead, 1959. Decommissioned under 1995–96 scheme.

Loyal Friend INS450, built by Macduff Boatbuilding & Engineering Co., Macduff, as *Loyal Friend* N280, 1976. Became *Loyal Friend* INS450, 1993; *Loyal Friend* FR606, 1996. Decommissioned 2011.

Wanderer III INS161, built by Campbeltown Shipyard Ltd, Campbeltown, as *Wanderer III* BA66, 1984. Became *Wanderer III* INS161, 1993. Moved to Eire and became *Owenie Padraig* S611, 2011.

Faithful INS38, built by James Noble (Fraserburgh) Ltd, Fraserburgh, 1970. Became *Faithful* WK126, 1995. Decommissioned 2002.

Gratitude BA130, built by Macduff Boatbuilding & Engineering Co., Macduff, as *Gratitude* INS272, 1968. Became *Gratitude* FY454, 1975; *Gratitude* SY54, 1985; *Gratitude* BA130, 1992; *Gratitude* BA830, 2000. Decommissioned 2002.

Huntress BA93, built by Alexander Noble & Sons Ltd, Girvan, as *Westerlea* OB93, 1971. Became *Huntress* BA93, 1980. Decommissioned 2002.

Weston Bay OB129, built by Cygnus Marine Ltd, Penryn, Cornwall, 1985.

Saffron BA172, built by Weatherhead & Blackie, Dunbar, 1973. Decommissioned 2002.

Girl May PD239, built by Thomas Summers & Co., Fraserburgh, as *Serene* LK63, 1955. Became *Serene* BF46, 1969; *Serene* SO647, 1976; *Scarlet Cord* B279, 1978; *Girl May* CN240, *c.*1979; *Girl May* PD239, 1988; *Girl May* PD788, 1996. Decommissioned 2000.

Silver Quest II BA122, built by J. & G. Forbes & Co., Sandhaven, as *Wistaria* FR116, 1972. Became *Silver Quest II* BA122, 1980. Sank west of Isle of Man, 30 September 2004.

Sea Otter BA6, built by Philip & Son Ltd, Dartmouth, as *Sea Otter* BM6, 1969. Became *Sea Otter* BA6, 1973. Decommissioned under 1995–96 scheme.

Shearbill GY571, built by Thomas Summers & Co., Fraserburgh, as *Shearbill* FD134, 1956. Became *Shearbill* GY571, 1958. Ceased fishing 1999 and converted to dive boat.

Oruna BA20, built by James N. Miller & Sons Ltd, St Monance, 1983. Decommissioned 2003.

Fair Morn V BA19, built by Macduff Boatbuilding & Engineering Co., Macduff, as *Athena* BF190, 1973. Became *Athena* CN262, 1986; *Athena* INS181, 1988; *Fair Morn V* BA19, 1991.

Lunaria AR82, built by Aas Mek Verksted A/S, Vestnes, Norway, as *Golden Star* B235, 1975. Became *Lunaria* AR82, 1984. Moved to Eire and became *Lunaria* SO896, 1996.

Stormdrift BA187, built by Alexander Noble & Sons Ltd, Girvan, 1962. Became *Stormdrift* CN368, 1994. Moved to Eire for conversion to houseboat, 2004.

Spes Bona III BA107, built by Gerrard Brothers, Arbroath, as *Lodestar* AH44, 1972. Became *Helann* OB434, 1988; *Spes Bona III* BA107, 1991; *Sheromhalidh* CN35, 1995; *Sapphire* CN697, 1999; *Sapphire II* SY108, 2004. Sank 6 miles east of Stornoway following collision, 12 January 2011.

Radiant Light N243, built by Jones Buckie Shipyard Ltd, Buckie, as *Crystal Sea* N243, 1967. Became *Radiant Light* N243, 1987. Decommissioned 2003.

Star of Peace II WK150, built by Jones Buckie Shipyard Ltd, Buckie, as *St Gerardine* INS251, 1957. Became *Star of Peace II* WK150, 1973. Decommissioned 1994.

Siskin AR302, built by Frederikssund Skibsvaerft, Frederikssund, Denmark, as *Biscaya* L658, 1969. Became *Paul Dam* GY436, 1979; *Carol Ann* GY436, 1980; *Wendy Louise* GY436, 1988; *Siskin* AR302, 1992. Moved to Maryport 1998. Sprang a leak and sank 10 miles east of St Bee's Head, Cumbria, 13 January 2005.

Breydon Mallard LT131, built by Richards (Shipbuilders) Ltd, Lowestoft, as *Boston Beaver* LT445, 1962. Became *Breydon Mallard* (oil rig standby vessel), 1978; *Breydon Mallard* LT131, 1987.

Flourish ME76, built by Thomas Summers & Co., Fraserburgh, as *Emulous* INS193, 1959. Became *Carbresia* A443, 1987; *Dorothy D III* ME76, 1988; *Flourish* ME76, 1993. Decommissioned 1996.

TROON

Britannia BA267, built by J. & G. Forbes & Co., Sandhaven, 1964. Became *Golden Dawn* BA267, 1994. Sold for conversion to pleasure boat 2000.

Andrew BA698, built by Jones Buckie Shipyard Ltd, Buckie, as *Olympic* INS16, 1957. Became *Olympic* UL16, 1965; *Olympic* LH303, 1982; *Andrew* LH303, *c.*1986; *Andrew* BA698, 1994. Ceased fishing 2001.

Brisca M16, built by James N. Miller & Sons Ltd, St Monance, 1988. Became *Brisca* UL37, 2009.

Spes Bona IV BA107, built by Alexander Noble & Sons Ltd, Girvan, as *Dawn Maid* CT99, 1969. Became *Dawn Maid* BA249, 1990; *Spes Bona IV* BA107, 1996; *Dawn Maid* TN102, 2004.

Silver Quest AR190, built by James Noble (Fraserburgh) Ltd, Fraserburgh, as *Eastern Dawn* FR82, 1971. Became *Bon Accord* BF55, 1985; *Leandris* WK471, 1991; *Incentive* BA471, 1993; *Myosotis* BA471, 1994; *Myosotis* AR190, 1995; *Silver Quest* AR190, 1996. Decommissioned 2022.

Constant Faith BA353, built by Macduff Engineering Co., Macduff, as *Blue Enterprise* LT488, 1964. Became *Orion* BF432, 1966; *Orion* LH332, 1974; *Constant Faith* FR303, 1980; *Constant Faith* BA353, 1988. Decommissioned 2002.

Ocean Maid BA55, built by J. & G. Forbes & Co., Sandhaven, as *Aubretia* B58, 1986. Became *Ocean Maid* BA55, 2020.

Trust AR871, built by Eyemouth Boatbuilding Co. Ltd, Eyemouth, as *Sanlormarho* LH220, 1974. Became *Sanlormarho* LK20, 1979; *Trust* INS46, 1988; *Trust* B46, 1991; *Trust* AR871, 1995. Constructive total loss after fire in 2015.

Pathfinder OB181, built by Alexander Noble & Sons Ltd, Girvan, as *Pathfinder* BA252, 1964. Became *Pathfinder* OB181, 1973.

Opportune BA221, built by Smith & Hutton (Boatbuilders) Ltd, Anstruther, as *Cairnsmore* BA74, 1970. Became *Cairnsmore* KY37, 1974; *Cairnsmore* OB318, 1979; *Marlaine* BA318, 1990; *Bonnie Doon* BA318, 1990; *Opportune* BA221, 1991; *Ocean Queen* RO91, 1995. Derelict at Port Bannatyne, Isle of Bute, from 2002 to 2012, when broken up.

Solstice II BF56, built by Ailsa Troon Ltd, Troon, 1998. Moved to Eire, 2003. (While *Solstice II* was for a time owned by a company registered in Eire, I can find no trace of her being registered as a fishing vessel there. I believe she was laid up in Killybegs for two years.) Moved to Orkney and became *Viking Monarch* K58, 2007. Moved to Dubai, 2014.

Atlantic Challenge PD197, built by Ailsa Troon Ltd, Troon, 1999.

ARDROSSAN

Bon Ami BA104, built by Chantier Naval Vergoz, Concarneau, France, as *Gwel Vo* LO614948, 1985. Became *Maluka* LO614948, 1999; *Maluka* FR372, 2015; *Bon Ami* BA104, 2019.

Menorca AR777, built by T. Mitchison Ltd, Gateshead, as *Japonica* A524, 1961. Became *Haponica* A524, 1983; *Menorca* AR777, 1992. Decommissioned 2013.

Islander BA316, built by Alexander Noble & Sons Ltd, Girvan, as *Islesman* SY433, 1967. Became *Islesman* BA141, 1976; *Boy Cameron II* BA141, 1977; *Boy Cameron II* INS314, 1983; *Islander* BA316, 1990.

ARDRISHAIG

Shemaron CN244, built by William Weatherhead, Cockenzie, as *Wistaria* BA64, 1949. Became *Shemaron* CN244, 1964. Ceased fishing 2006. Preserved by Ring Net Heritage Trust.

TARBERT

Caledonia TT34, built by Whitby Shipyard, Whitby, as *Success* LH81, 1975. Became *Success* WY212, *c.*1986; *Deliverer* WY212, 2001; *Caledonia* TT34, 2003.

Frigate Bird TT137, built by Macduff Boatbuilding & Engineering Co., Macduff, as *Loranthus* BCK133, 1977. Became *Lodestar* AH54, 1988; *Shannon* FR4, 1996; *Frigate Bird* TT137, 2002.

Destiny TT279, built by Gerrard Brothers, Arbroath, as *Stella Maris III* CN250, 1974. Became *Stella Maris III* CY250, 1983; *Destiny* TT279, 2008.

CARRADALE

Village Belle IV TT74, built by Alexander Noble & Sons Ltd, Girvan, 1970. Became *Village Belle IV* B157, 2011; *Village Belle IV* B377, 2013; reverted to *Village Belle IV* TT74, 2015.

CAMPBELTOWN

Adoration II CN78, built by Fairlie Yacht Slip, Fairlie, as *Adoration II* BA238, 1964. Became *Adoration II* CN78, 1972.

Boy Andrew TT179, built by Gerrard Brothers, Arbroath, as *Carona* KY179, 1973. Became *Boy Andrew* TT179, 1990.

TAYINLOAN

Louise Joanne SS106, built by G. Percy Mitchell, Portmellon, near Mevagissey, Cornwall, 1974. Moved to Eire and became *Louise Joanne* WT252, 2003.

MALLAIG

Vikingborg OB285, built by Herd & Mackenzie, Buckie, as *Vikingborg* H285, 1975. Became *Vikingborg* OB285, 2012; *Vikingborg* SY275, 2017.

Spes Nova CN77, built by Alexander Noble & Sons Ltd, Girvan, as *Spes Bona II* BA107, 1972. Became *Spes Nova* CN77, 1991. Sank north-east of Portree, Skye, 30 November 2007.

Contest LK70, built by J. & G. Forbes & Co., Sandhaven, 1985.

Mareather OB503, built by Herd & Mackenzie Ltd, Buckie, as *Unity II* BCK35, 1986. Became *Caledonia* BCK35, 1993; *Aubretia* BCK32, 2007; *Mareather* OB503, 2016.

Caralisa OB956, built by James N. Miller & Sons Ltd, St Monance, as *Constant Hope II* KY100, 1975. Became *Ardgour VI* KY100, 1996; *Caralisa* OB956, 2000; *Strathyre* OB885, 2019; *Strathyre* N85, 2019.

Hercules II LK438, built by Alexander Noble & Sons Ltd, Girvan, as *Hercules II* BA7, 1968. Became *Hercules II* UL156, 1974; *Hercules II* LK438, 1981. Decommissioned 1995.

GAIRLOCHY

Silver Cloud II WK80, built by Gerrard Brothers, Arbroath, 1972.

EYEMOUTH

Maryeared TT57, built by Weatherhead & Blackie, Port Seton, 1965. Converted to pleasure boat 2000.

Immanuel VII LH546, built by MacLean & MacRae, Kyle of Lochalsh, as *Amazon* BRD241, 1980 (fitting out of GRP hull built by Halmatic (Scotland) Ltd, Kirkwall, *c.*1973). Became *Amazon* PD271, 1985; *Amazon* K271, 1990; *Amazon* LK271, 2007; *Immanuel VII* B78, 2008; *Immanuel VII* LH546, 2015; *Julia Anne* FD5, 2018.

L'Ogien K62, built by Ateliers et Chantiers Breheret, Ingrandes-Le Fresne sur Loire, France, as *L'Ogien*, 1979. Became *Simbad*, 1983; *Celtic Rose* S264, 1997; *Chelaris J* GU323, 2000. Sank 6 miles south of Alderney, 1 October 2003. Salvaged 21 November 2003. Became *L'Ogien* PZ28 by 2011; *L'Ogien* K62, 2014.

PITTENWEEM

St Adrian II KY202, built by D.A.D. Munro Engineering & Contractors Ltd, Old Kilpatrick, 1981. Vessel delivered incomplete for final completion by owner. Became *Imogen II* PZ110, 2010; *Imogen II* FY470, 2011; *Kayleigh M* OB15, 2012.

GOURDON

Ocean Vanguard OB21, built by James Noble (Fraserburgh) Ltd, Fraserburgh, as *Ocean Vanguard* BH3, 1954. Became *Ocean Vanguard* KY131, 1979; *Ocean Vanguard* OB21, 1993; *Vivid* ME106, 1999.

ABERDEEN

Ross Anne A155, built by Campbeltown Shipyard Ltd, Campbeltown, as *Shielwood* A155, 1977. Became *Ross Anne* A155, 1997. Decommissioned 2002.

Alexanders A177, built by J. & G. Forbes & Co., Sandhaven, 1972. Became *Primrose* FR223, 1999. Decommissioned under 2003–04 scheme.

Armana FD322, built by Drypool Engineering & Dry Dock Co. Ltd, Hull, 1976. Moved to Eire and became *Armana* D327, 1994. Scrapped 1996.

Radiant PD298, built by Astilleros La Parrilla SA, Asturias, Spain, 2001. Capsized and sank after fouling fishing gear on seabed obstruction 45 miles north-west of Isle of Lewis, 10 April 2002.

Argonaut IV KY157, built by Campbeltown Shipyard Ltd, Campbeltown, 1976. Moved to Eire and became *Argonaut IV* DA22, 2003.

Arktos II KY129, built by Campbeltown Shipyard Ltd, Campbeltown, 1979. Sank 75 miles west of Stavanger, Norway, 1 March 2003.

Natalie B H1074, built by Maaskant Machinefabrieken, Bruinisse, Netherlands, as *Klaase* GO7, 1967. Became *De Twee Gebroeders* OD7, 1974; *Marrie Jacob* D141, 1981; *Natalie B* BA238, 1986; *Natalie B* H1074, 1996; *Natalie B* A107, 2012; *Natalie C* A108, 2016; *KMS Kinnaird*, 2017. Now a guardship for offshore installations.

St Amant BA101, built by John Harker (Shipyards) Ltd, Knottingley, as *Cassamanda* SH38, 1975. Became *St Amant* SH38, 1978; *St Amant* BA101, 1986; *Shannon Kimberly* RY169, 2014.

Academus BA817, built by Hepworth Shipyard Ltd, Paull, Hull, 1998.

FRASERBURGH

Britannia BF178, built by George Thomson & Sons, Buckie, as *Britannia* INS68, 1952. Became *Britannia* BF178, 1974. Decommissioned under 1995–96 scheme. Broken up 2020.

Serenity BF24, built by Macduff Shipyards Ltd, Macduff, 1991.

Silver Fern FR416, built by Herd & Mackenzie, Buckie, 1982. Became *Silver Fern* CY197, 2018.

Our Hazel UL543, built by Astilleros Armon SA, Navia, Spain, as *Our Hazel* GU171, 1994. Became *Our Hazel* UL543, 2000.

Conquest FR1, built by Hugh McLean & Sons Ltd, Renfrew, 1968. Decommissioned 1998.

Isla S DS1, built by Maaskant Machinefabrieken, Stellendam, Netherlands, as *Vrouw Grietje* GO14, 1990. Became *Andries de Vries* UK143, 2005; *Isla S* DS1, 2011.

Resilient FR327, built by James N. Miller & Sons Ltd, St Monance, as *Vestrfjordr* K274, 1986. Became *Resilient* FR327, 1998. Sank 37 miles south of Sumburgh, Shetland, 11 February 2001.

Verwachting P845, built by Van Goor's Scheepsbouw & Machinefabrieken, Monnickendam, Netherlands, as *Verwachting* UK176, 1969. Became *Verwachting* P845, 1982.

Kimara FR176, built by J. & G. Forbes & Co., Sandhaven, 1975. Decommissioned 2002.

Fladda Maid UL209, built by Macduff Boatbuilding & Engineering Co., Macduff, as *Endeavour* BF326, 1989. Became *Enterprise* BF326, 1993; *Good Design* BF151, 1999; *Crystal Sea* SS118, 2007; *Fladda Maid* UL209, 2016; *Aspire* B903, 2021.

Fredwood II BA340, built by Johs Kristensen Skibsbyggeri ApS, Hvide Sande, Denmark, 1981. Ceased fishing 2006.

Orion BF432, built by Macduff Shipyards Ltd, Macduff, as *Amethyst II* BF19, 2000. Became *Virtuous* FR253, 2003; *Orion* BF432, 2010; *Faithful Friend* FR83, 2021.

Kings Cross FR380, built by Eidsvik AS, Uskedalen, Norway, 2003. Moved to Norway and became *Aakerøy* N-200-DA, 2014.

Forever Grateful FR249, built by Astilleros Zamakona SA, Santurce, Bilbao, Spain, 2001. Moved to Norway and became *Trygvason* H-20-B, 2016.

Georgia Dawn INS140, built by Macduff Shipyards Ltd, Macduff, 2005.

Mia Jane W FR443, built by Chantier Piriou Frères, Concarneau, France, as *Nicolas Jeremy* BL734995, 1990. Became *Boy Jason* S41, 2001; *Ocean Reaper II* S41, 2003; *Mia Jane W* FR443, 2008; *Boy Conor* N443, 2022.

Provider II BF422, built by Miller Methil Ltd, Methil, as *Providence II* BF422, 2001. Became *Provider II* BF422, 2005; *Provider II* BM422, 2012.

Fertile FR599, built by Sigbjorn Iversen Mek Verksted, Flekkefjord, Norway, as *Starcrest* PD232, 1989. Became *Ocean Surf* FR225, 1991; *Fertile* FR599, 1995; *Havilah* N200, 2002. Moved to Faroe and became *Fram* VN449, 2014.

Kinnaird FR377, built by Stocznia Remontowa Parnica Sp.z.o.o., Parnica, Poland, 1996. Decommissioned 2003.

Coronata II BF356, built by John Lewis & Sons Ltd, Montrose, as *Donside* A552, 1961. Became *Donside* BCK234, 1966; *Adastra* BCK234, 1973; *Culblean* BCK234, 1976; *Culblean* BA74, 1977; *Misty Isle II* BA74, 1978; *Coronata II* BF356, 1980. Decommissioned 2004.

Modulus FR272. Hull built by Ryton Marine Ltd, Wallsend. Yard suffered financial collapse before vessel completed. Vessel fitted out by British United Trawlers Engineering (Grimsby) Ltd, Grimsby, 1975, after abortive attempts to have her completed at Whitby Shipyard and Berwick Shipyard. Originally named *Andree* PD161. Delivered as *Andree* BCK160. Became *Modulus* FR272, 1978; *Defiance* FR385, 1998. Ceased fishing 2019. Now a guardship for offshore installations.

Suilven IV BCK366, built by Macduff Boatbuilding & Engineering Co., Macduff, as *Chelaris* BF16, 1984. Became *Arctic Gull* DA4, 1990; *Suilven IV* BCK366, 1994. Decommissioned 2010.

Daystar BF151, built by Macduff Shipyards Ltd, Macduff, 2007. Became *Zenith* BF106, 2016; *Unity* B106, 2022.

Ocean Harvest BF145. Hull built by Ryton Marine Ltd, Wallsend. Yard suffered financial collapse before vessel completed. Incomplete hull designated *Olivia* PD162 fitted out by British United Trawlers Engineering (Grimsby) Ltd, Grimsby, and delivered as *Troilus* BCK159, 1975, after abortive attempts to have her completed at Whitby Shipyard and Berwick Shipyard. Became *Ocean Harvest* BF145, 1978. Decommissioned 2002.

Zara Annabel BCK126, built by Kramer & Booy NV, Kootstertille, Netherlands, as *Johannes Post* UK57, 1968. Became *Zeven Gebroeders* UK236, 1973; *Nellie* VD54, 1979; *Nellie* SU143, 1987; *Nellie* PZ10, 1998; *Calisha* PD235, 2006; *Zara Annabel* BCK126, 2018.

Quantus PD379, built by Eidsvik AS, Uskedalen, Norway, 2008.

Pindarus FR159, built by Cubow Ltd, Woolwich, as *Unity* PD209, 1975. Became *Constant Hope* PD209, 1985; *Paramount* PD209, 1988; *Radiant Star II* PD251, 1993; *Radiant Star II* FR159, 1997; *Pindarus* FR159, 1997. Decommissioned 2002.

Bracoden BF37, built by Macduff Shipyards Ltd, Macduff, as *Solstice* BF56, 1991. Became *Solstice* LK272, 1998; *Bracoden* BF37, 1998; *Osha* CY141, 2020; *Solstice* BF56, 2022.

Davanlin FR890, built by Macduff Shipyards Ltd, Macduff, as *Achieve* BF223, 1995. Became *Rebecca* FR143, 2001; *Davanlin* FR890, 2008.
Replenish BF28, built by Macduff Shipyards Ltd, Macduff, 1998. Became *Opportunus* PD965, 2018. Moved to Belize 2019.
Carvida FR457, built by James N. Miller & Sons Ltd, St Monance, 1988. Became *Seaward Quest* BF377, 1996; *Phoenix* FR941, 2001.
Winner FR265, built by Macduff Engineering Co., Macduff, as *Winner* LK149, 1956. Became *Winner* FR265, *c.*1963. Ceased fishing 1996.
Aquinis BA500, built by Hepworth Shipyard Ltd, Paull, Hull, 1992. Became *Aquinis* BM519, 2016; *Henry Monty* BM2, 2019.
Sunbeam FR487, built by Astilleros Zamakona SA, Santurce, Bilbao, Spain, 1999.
Grateful FR249, built by Karstensens Skibsvaerft AS, Skagen, Denmark, 2017.

MACDUFF

Audacious BF83, built by Jones Buckie Shipyard Ltd, Buckie, as *Crystal River* BCK16, 1985. Became *Audacious* BF83, 1991; *Endurance III* FR98, 1995; *Good Hope* FR891, 1999. Decommissioned 2004.
Intrepid BA100, built by George Thomson & Sons, Buckie, 1970.

BUCKIE

Nordic Prince BCK18, built by Fiskerstrand Verft A/S, Fiskerstrand, Norway, as *Joffre* M-8-G, 1975. Became *Sandevaering* M-82-S, 1985; *Lestaskjaer* M-92-G, 1991; *Suderøy* M-S-72, 1999; *Nordic Prince* BCK18, 2000. Decommissioned 2003.

LOSSIEMOUTH

Clarness INS51, built by Gerrard Brothers, Arbroath, as *Ardgour IV* KY107, 1987. Became *Clarness* INS51, 2000; *Lindisfarne* INS51, 2008. Decommissioned 2010.

BURGHEAD

Alex Watt INS163, built by John Watt & Sons, Banff, as *Alex Watt* BF362, 1966. Became *Alex Watt* INS163, 1976. Wrecked at Skye, 24 January 2000.

PETERHEAD

Accord PD90, built by Mylebust Mek Verksted AS, Gurskebotn, Norway, as *Altaire* LK429, 1979. Became *Serene* LK297, 1987; *Accord* PD90, 1995. Moved to Denmark and became *Grenen* S108, 2004.

Della Strada CY158, built by Maaskant Machinefabrieken, Stellendam, Netherlands, as *Heritage* BF150, 1972. Became *Ambassador* BF450, 1994; *Della Strada* CY158, 1995. Moved to Africa 2003, later to Finland and became *Vanderlax*.

Allegiance SH90, built by Cochrane Shipbuilders Ltd, Selby, 1987.

Tino GY203, built by Glyngøre Skibs-og Baadebyggeri, Glyngøre, Denmark, 1972. Ceased fishing *c.* 2001. Converted to houseboat.

Favonius PD17, built by Fairmile Construction Co. Ltd, Berwick-upon-Tweed, 1969. Decommissioned 2002.

Starlight PD150, built by Cubow Ltd, Woolwich, 1975. Decommissioned 2002.

Serene PD58, built by James Noble (Fraserburgh) Ltd, Fraserburgh, 1974. Decommissioned 1998.

Barbarella PD134, built by Bideford Shipyard Ltd, Bideford, as *Barbarella* E30, 1969. Became *Barbarella* PD134, 1974. Moved to Eire and became *Barbarella* SO960, 2000.

Faithful FR129, built by J. & G. Forbes & Co., Sandhaven, 1985. Became *Faithful II* FR429, 2020 (original name and number wanted for transfer to a new vessel); *Osprey* FR78, 2021.

Evening Star PD1022, built by Hepworth Shipyard Ltd, Paull, Hull, as *Tobrach N* TN2, 1991. Became *Evening Star* BA841, 2005; *Evening Star* PD1022, 2006.

Cardanel BCK9, built by J. Samuel White & Co. (Scotland) Ltd, Cockenzie, as *Provider* SY24, 1968. Became *Notre Dame* CY89, 1974; *Notre Dame* BCK9, 1983; *Cardanel* BCK9, 1988. Sank north-east of Cromer in Norfolk, 13 September 1998.

Silvery Sea OB245, built by Maaskant Machinefabrieken, Stellendam, Netherlands, 1976. Sank after collision with freighter 30 miles west of Esbjerg, Denmark, 14 June 1998.

Sunbeam FR478 built by Flekkefjord Slipp & Maskinfabrikk, Flekkefjord, Norway, as *Abba* UA960, 1975. Became *Gerda Marie* H-32-AV, 1976; *Sunbeam* FR478, 1987; *Sunbeam II* FR478, 1999; *Sunbeam II* (no fishing registration), 1999. Moved to Bolivian registry retaining name *Sunbeam II*, 2004.

Iolanthe PD131, built by Hugh McLean & Sons Ltd, Renfrew, as *Replenish* BF84, 1971. Became *Maranatha* FR291, 1978; *Iolanthe* PD131, 1993. Sank east of Shetland, 5 September 2001.

Lapwing PD972, built by Campbeltown Shipyard Ltd, Campbeltown, as *Xmas Rose* FR125, 1973. Became *Defiance* FR125, 1989; *Defiance* FR385, 1990; *Lapwing* PD972, 1998. Ceased fishing and became guardship for offshore installations 2020.

Ajax INS168, built by Campbeltown Shipyard Ltd, Campbeltown, 1975. Decommissioned 2003.

Starina INS801, built by Richard Dunston (Hessle) Ltd, Thorne, as *Starina* LK347, 1976. Became *Starina* INS801, 1997. Decommissioned 2003.

Advance II INS77, built by Campbeltown Shipyard Ltd, Campbeltown, as *Brereton* INS97, 1987. Became *Valhalla IV* LH184, 1989; *Advance II* INS77, 1999; *Advance* WY77, 2011; *Ardent* INS127, 2016.

Sette Mari LK257, built by Götaverken Arendal AB, Arendal, Gothenburg, Sweden, as *Sette Mari* GG59, 1987. Became *Sette Mari* LK257, 1992. Moved to Norway and became *Sette Mari* SF-7-A, 1997.

Vertrauen BF450, built by Maritem Industries Ltd, Cobh, Co. Cork, Eire, as *Girl Stephanie* G190, 1982. Became *Vertrauen* BF450, 1996. Sprang a leak and sank 80 miles north-east of Fraserburgh, 19 July 2001.

Provider PD250, built by J. & G. Forbes & Co., Sandhaven, as *Rose of Sharon III* LH56, 1983. Became *Minerva* FR147, 1991; *Provider* PD250, 1998. Decommissioned 2002.

Ocean Bounty PD182, built by Macduff Shipyards Ltd, Macduff, 1998.

Fairwind FR317, built by Hugh McLean & Sons Ltd, Renfrew, as *Wave Crest* LK276, 1969. Became *Wave Crest* BF105, 1986; *Stephens* FR156, 1987; *Fairwind* PD197, 1990; *Fairwind* FR317, 1991. Decommissioned 1998.

Alert FR396, built by Sigbjorn Iversen Mek Verksted, Flekkefjord, Norway, as *Charisma* LK362, 1979. Became *Alert* FR396, 1995; *Charisma* CY88, 2002. Decommissioned *c.*2009.

Morning Dawn PD359, built by Maaskant Machinefabrieken, Bruinisse, Netherlands, as *Pieter Jr* HD38, 1992. Became *Morning Dawn* PD359, 1998; *Our Anna* PD657, 2005; *Our Anna* PZ657, 2012.

Fruitful Harvest III PD247, built by James Noble (Fraserburgh) Ltd, Fraserburgh, 1976. Became *Fruitful Harvest III* PD199, 2008; *Fruitful Harvest III* GY991, 2008. Sold for conversion to pleasure boat 2017.

Jasper III PD174, built by John Lewis & Sons Ltd, Aberdeen, as *Pisces* A193, 1971. Became *Jasper III* PD174, 1993. Sank 60 miles east of Orkney, 10 September 1999.

Silver Star BCK131, built by Herd & Mackenzie, Buckie, as *Unity* BCK57, 1972. Became *Silver Star* FR73, 1986; *Silver Star* BCK131, 1988. Decommissioned 2002.

Wave Crest BCK217, built by Herd & Mackenzie, Buckie, 1974. Became *Ocean Way* FR349, 2006. Sank 100 miles east of Farne Islands, 2 November 2015.

Arkh Angell K616, built by Campbeltown Shipyard Ltd, Campbeltown, 1990. Moved to Eire and became *Arkh-Angell* DA33, 2011.

Convallaria VI BF58, built by Maaskant Machinefabrieken, Stellendam, Netherlands, as *Andra Tait* FR226, 1978. Became *Convallaria VI* BF58, 1997. Moved to Iran 2004.

Cap Saint Georges BL924675, built by Chantier Piriou Frères, Concarneau, France, 2003.

Diligent PD314. Hull built by George Brown & Co. (Marine) Ltd, Greenock, under sub-contract to Smith & Hutton (Boatbuilders) Ltd, Anstruther. Following receivership of Smith & Hutton completed by J. & G. Forbes & Co., Sandhaven, as *Coronella* BF277, 1977. Became *Linarolynn* FR362, 1980; *Diligent* FR362, 1986; *Diligent* PD314, 1988. Moved to Sweden and became *Hållö* LL149, 1999.

Sparkling Star PD137, built by K Hakvoort Scheepsbouw NV, Monnickendam, Netherlands, 1975. Moved to Eire and became *Sparkling Star* D437, 2000.

Resplendent PD225, built by Astilleros La Parrilla SA, Asturias, Spain, 2000. Moved to Namibia 2003. Became *Resplendent* L1190. Sank 18 February 2020.

Boy Ken II DO5, built by George Forbes (Peterhead) Ltd, Peterhead, as *Maureen* CN185, 1949. Became *Boy Ken The Second* B23, 1954; *Boy Ken II* DO5, 1986.

Atlantic Challenge PD197, built by Ailsa Troon Ltd, Troon, 1999.

Krossfjord BF70, built by Skaalurens Skipsbyggeri, Rosendal, Norway, as *Krossfjord* H-69-S, 1964. Became Krossfjord BF70, 1984. Moved to Norway and became *Nordsnurp* N-21-VV, 1997.

Taits FR228, built by Karmøy Mek Verksted, Kopervik, Norway, 1978. Became *Taits II* FR229, 2001. Moved to Norway and became *Leik II* R-44-K, 2002.

Resolute BF50, built by West Contractors AS, Ølensvaag, Norway, 2003. Became *Artemis* BF60, 2020. Moved to Denmark 2021.

Eilean Croine CY190, built by Baltimore Boatyard Co. Ltd, Baltimore, Eire, as *Oilean Croine* SO690, 1979. (Hull built by K Hakvoort Scheepsbouw NV, Monnickendam, Netherlands.) Became *Eilean Croine* CY190, *c.*1989. Moved to Eire and became *Eilean Croine* S238, 1998.

Shemara PD314, built by Maritem Industries, Cobh, Co. Cork, Eire, as *Brendelen* SO709, 1980. (Hull built by Scheepswerf de Amstel BV, Ouderkerk aan de Amstel, Netherlands.) Became *Brendelen* FR709, 1995; *Shemara* FR124, 1996; *Shemara* PD314, 1999. Moved to Finland 2006.

Pathway PD165, built by Simek AS, Flekkefjord, Norway, 1998. Became *Pathway I* PD65, 2003. Moved to Faroe and became *Jupiter* FD125, 2004. Moved to Norway and became *Aakerøy* N-300-DA, 2007.

Pathway PD165, built by Karstensens Skibsvaerft AS, Skagen, Denmark, 2017.

Radek H-15-AV, built by Bolsønes Verft AS, Molde, Norway, as *Liaholm*, 1976. Became *Øvrabøen* R-5-B, 1981; *Noragutt* N-45-LN, 2002; *Gangstad Junior* M-5-MD, 2003; *Radek* H-15-AV, 2006.

Jubilee Spirit GY25, built by Hepworth Shipyard Ltd, Paull, Hull, as *Jubilee Quest* GY900, 1997. Became *Jubilee Spirit* GY25, 2009.

Lomur LK801, built by Trønderverftet A/S, Hommelvik, Norway, as *Rósa* HU294, 1988. Main construction work by Johan Drage A/S, Rognan, Norway. Became *Budafell* SU90, 1991; *Lomur* HF177, *c.*1993; *Lomur* K177, 1994; *Lomur* LK801, 1996. Constructive total loss after flooding in Buckie harbour. Repaired and sold to Malta, where she reverted to the name *Budafell*, 2002.

Wiron 5 PH1100, built by Construcciones Navales P Freire SA, Vigo, Spain, as *Wiron 5* SCH22, 2002. Became *Wiron 5* PH1100, 2014.

Harvester PD98, built by Alexander Noble & Sons Ltd, Girvan, as *Devorgilla* BA67, 1978. Became *Devorgilla* LH217, 1982; *Harvester* PD98, 1985; *Good Hope* FR891, 1996; *Good Hope* B900, 1998. Decommissioned 2002.

Ocean Harvest PD198, built by Sigbjorn Iversen Mek Verksted, Flekkefjord, Norway, as *Summer Dawn* PD64, 1973. Became *Ocean Harvest* PD198, 1996; *Summer Dawn* PD97, 2008; *Adventure* (no fishing registration) and switched to offshore guardship work, 2014.

Vigilant PD365, built by Sigbjorn Iversen Mek Verksted, Flekkefjord, Norway, 1980. Moved to Iceland and became *Jóna Edvalds* SF20, 1995.

Vigilant PD365, built by Simek AS, Flekkefjord, Norway, 1995. Moved to Norway and became *Aakerøy* N-200-DA, 2002; *Herøy* M-201-HØ, 2007. Sank after running aground near Kristiansund, 25 January 2007.

Glenugie PD347, built by Jones Buckie Shipyard Ltd, Buckie, as *Prestige* INS305, 1981. Became *Glenugie* PD347, 1988; *Brighter Dawn* LK154, 1999; *Surina* BCK95, 2002; *Tranquility* LK63, 2008; *Fortuna* LK29, 2011; *Resolution* WY78, 2012; *Resolution* FR78, 2016.

Magnificent PD250, built by Richards (Shipbuilders) Ltd, Lowestoft, as *Ethel Mary* LT337, 1957. Became *Golden Promise* FR186, 1970; *Golden Promise* PD250, 1979; *Magnificent* PD250, 1979. Decommissioned 1995.

Sea Lady TN20, built by Scheepswerf & Constructiebedrijf Marcon, Hoogezand, Netherlands, as *Fokke Grietje* UK126, 1986. Became *Andrea* GY17, 1991; *Andrea* PD962, 1997; *Lady T Emile* BM2000, 2000; *Sea Lady* BM2000, 2010; *Sea Lady* TN20, 2011; *Sea Lady* BM200, 2021; *Stella Maris* BM200, 2022.

Leanne PD345, built by Campbeltown Shipyard Ltd, Campbeltown, as *Benaiah II* B350, 1981. Became *Discovery* BF268, 1994; *Leanne* PD345, 2016.

Scotia PD233, built by Bangor Shipyard Ltd, Bangor, Co. Down, Northern Ireland, as *Fragrant Cloud II* B430, 1980. Became *Fragrant Cloud II* PD233, 1984; *Scotia* PD233, 1994. Decommissioned 2003; sold for conversion to houseboat instead of scrapping.

Sustain PD378, built by John R. Hepworth (Hull) Ltd, Paull, Hull, as *Morning Dawn* PD195, 1975. Became *Lupina* PD495, 1992; *Sustain* PD378, 1992. Decommissioned under 2003–04 scheme.

Altaire LK429, built by Solstrand AS, Tomrefjord, Norway, 2004.

Aalskere K373, built by Stocznia Remontowa Parnica Sp.z.o.o., Parnica, Poland, as *Vandal* LK337, 1997. Became *Aalskere* K373, 2000; *Gemma Jane* K184, 2019. Sold to international interests 2020.

Fear Not II PD354, built by Gerrard Brothers, Arbroath, as *Sea Spray II* PD245, 1986. Became *Fear Not II* PD354, 2002; *Fear Not II* CN354, 2014.

Courage BF212, built by Smedvik Mek Verksted, Nordfjord, Norway, as *Azalea* LK396, 1980. Became *Courage* BF212, 1997. Moved to Faroe and became *Atlantsfarid* VA218, 2003.

Prowess BF720, built by K Hakvoort Scheepsbouw NV, Monnickendam, Netherlands, as *Paula* SO720, 1980. Became *Avril* (no fishing registration), 1994; *Prowess* BF720, 1995; *Prowess* CY720, 2003. Moved to Iceland and became *Que Sera Sera* HF26, 2006.

Moremma PD135, built by Jones Buckie Shipyard Ltd, Buckie, as *Moremma* BCK135, 1988. Became *Moremma* PD135, 1998.

Sunrise FR359, built by Campbeltown Shipyard Ltd, Campbeltown, 1984.

Shalimar PD303, built by John Lewis & Sons Ltd, Aberdeen, as *Salamis* PD142, 1974. Became *Harvest Reaper* PD142, 1983; *Shalimar* PD303, 1994. Decommissioned 2002.

Elysian PD444, built by George Thomson & Sons, Buckie, as *Elysian* LH375, 1960. Became *Elysian* LK444, 1974; *Elysian* PD444, 1987; *Immanuel II* N78, 1997; *Elysian* N912, 1997. Decommissioned 2004.

Renown FR246, built by McCrindle (Shipbuilding) Ltd, Ardrossan, 1987.

Westward BF350, built by Sterkoder Mek Verksted AS, Kristiansund, Norway, as *Sheanne* SO716, 1980. Became *Radiant Star* BF77, 1994; *Westward* BF350, 1996. Moved to Faroe and became *Hallarklettur* TN1161, 2002.

Nordfjordr PD118, built by James N. Miller & Sons Ltd, St Monance, as *Nordfjordr* K139, 1989. Became *Nordfjordr* PD118, 1997; *Audacious* BF83, 2013; *Nordfjordr* BF830, 2018; *Defiant*, 2018. Now a guardship for offshore installations. Note: all available sources give the name of this vessel as *Nordfjordr*. The name actually on the vessel was *Nørdfjørdr*.

Vandijck BM362, built by Scheepswerven Leon de Graeve, Zeebrugge, Belgium, as *Vandijck* Z162, 1974. Became *Vandijck* BM362, 1990.

Success PD789, built by Thames Launch Works Ltd, Twickenham, as *Gleaner* CN284, 1967. Became *Golden Years* LH14, 1969; *Golden Years* TN8, 1991; *Golden Years* DS10, 1994; *Success* PD789, 1996. Derelict on rocks at Leverburgh Pier, Isle of Harris, after breaking away from her berth in hurricane force winds on 11 January 2005.

Bonaventure LH111, built by Herd & Mackenzie Ltd, Buckie, 1987. Became *Bonaventure* BCK262, 2016; *Elkanah* BF200, 2019.

Heatherbelle VI LH272, built by Herd & Mackenzie Ltd, Buckie, as *Rebecca* LH11, 1987. Became *Heatherbelle VI* LH272, 1994; *Rebecca* LH11, 1997; *Sapphire IV* CN355, 2015; *Sapphire IV* N35, 2019.

Sharon's Rose LH317, built by Jones Buckie Shipyard Ltd, Buckie, as *Telstar* LK378, 1988. Became *Sharon's Rose* LH317, 1995; *Betty James* INS279, 2000. Sank after running aground on Island of Rhum, 10 July 2000.

Viking Warrior PD188, built by Nexo Skibs-og Baadebyggeri, Nexo, Bornholm, Denmark, as *Viking Warrior* FE183, 1968. Became *Viking Warrior* PD188, 1975; *Viking Warrior* BS188, 2012.

Ocean Way PD465, built by Sigbjorn Iversen Mek Verksted, Flekkefjord, Norway, 1989. Became *Torbas* F-7-M, 1997; *Unity* FR165, 2001. Moved to Norway 2013 and became *Staalegg* M-10-S, 2014.

Scottish Maid B417, built by J. & G. Forbes & Co., Sandhaven, as *Scottish Maid* BF317, 1979. Became *Scottish Maid* B417, 1994; *Amber Rose* B417, 1998. Sank off Isle of Man, 15 October 1998.

Opportunus PD96, built by Herd & Mackenzie Ltd, Buckie, as *Ajax* INS168, 1968. Became *Harvest Hope* PD96, 1972; *Star of Bethlehem* PD96, 1975; *Opportunus* PD96, 1987; *Benaiah III* N841, 1995. Decommissioned 1997.

Kings Cross PD365, built by Karstensens Skibsvaerft AS, Skagen, Denmark, 2016. Moved to Norway and became *Gerda Marie* H-365-AV, 2019.

Ocean Star FR894, built by Sigbjorn Iversen Mek Verksted, Flekkefjord, Norway, as *Challenge* FR77, 1971. Became *Ocean Star* FR894, 1996; *Sparkling Star* PD137, 2000. Moved to Sweden and became *Sette Mari* GG59, 2004.

Fair Morn PD224, built by Thomas Summers & Co., Fraserburgh, as *Girl Jean* FR92, 1959. Became *Fair Morn* PD224; *Fair Morn* FH29.

Christina S FR224, built by Ulstein Hatlø, Ulsteinvik, Norway, as *Ny Dolsøy* M-22-VD, 1978. Became *Ny Dolsøy* FR224, 1987; *Christina S* FR224, 1988. Moved to Norway and became *Rogne* M-22-HØ, 1997.

Christina S FR224, built by Flekkefjord Slipp & Maskinfabrikk, Flekkefjord, Norway, 1997. Moved to Denmark and became *Beinur* HG62, 2006.

Fair Morn INS308, built by Astilleros Armon SA, Navia, Spain, 1997. Became *Fair Morn* PD67, 2000; *Maracestina* INS291, 2008; *Maracestina* N291, 2016.

Venture PD72, built by Richard Dunston (Hessle) Ltd, Hessle, 1989. Became *Venturous* PD72, 1999; *Venturous* LK75, 1999; *Angelina* LK377, 2019.

Ryanwood FR307, built by Jones Buckie Shipyard Ltd, Buckie, as *Faithful* UL179, 1988. Became *Genesis* BF505, 1994; *Lorena II* BF811, 1998; *Ryanwood* FR307, 2002; *Summer Dawn II* PD97, 2014.

Regina Caeli CY58, built by Maaskant Machinefabrieken, Stellendam, Netherlands, as *Convallaria V* BF58, 1974. Became *Regina Caeli* CY58, 1998. Moved to Africa 2003, later to Finland.

Chris Andra FR228, built by West Contractors AS, Ølensvaag, Norway, 2006.

Boy John INS110, built by Macduff Shipyards Ltd, Macduff, 1996. Became *Wee Boy John* INS15, 2014 (original name and number wanted for transfer to a new vessel); *Karen of Ladram* PW3, 2014.

Ventura INS334, built by Richard Irvin & Sons Ltd, Peterhead, as *Graceful* PD133, 1974. Became *Graceful* FR147, 1988; *Tranquility* INS35, 1990; *Ventura* INS334, 1994. Ceased fishing by 1998. Moved to Honduras 1999.

Paragon V PD786. Partially built by Jones Buckie Shipyard Ltd, Buckie, and completed by Dunston (Ship Repairers) Ltd, Hull, 1995. Became *Starlight* PD786, 2002; *Floreat* SH60, 2014.

Ability PD981, built by John Lewis & Sons Ltd, Aberdeen, as *Supreme* A476, 1976. Became *Supreme* INS276, 1981; *Adorne* INS220, 1983; *Ability* PD981, 2000. Decommissioned 2002.

Voyager N905, built by Flekkefjord Slipp & Maskinfabrikk, Flekkefjord, Norway, 1997. Moved to Faroe and became *Naeraberg* KG14, 2008.

Strathnairn INS185, built by J. & G. Forbes & Co., Sandhaven, as *Janet Sate* FR107, 1984. Became *Strathnairn* INS185, 1990; *Alba* PD181, 1998; *Carisanne II* FR951, 2002. Decommissioned 2011.

Falcon H119, built by James N. Miller & Sons Ltd, St Monance, as *Inter Nos II* KY57, 1985. Became *Falcon* FR119, 1996; *Falcon* H119, 2013.

Fisher Rose FR896, built by Campbeltown Shipyard Ltd, Campbeltown, as *Valhalla III* LH67, 1982. Became *Fisher Rose II* LH67, 1989; *Fisher Rose II* FR896, 1997; *Fisher Rose* FR896, 1997. Decommissioned 2002.

Our Lass III WY261, built by Parkol Marine Engineering Ltd, Whitby, 2013.

Victory Rose WY34, built by Parkol Marine Engineering Ltd, Whitby, 2017.

Serene FR491, built by Sigbjorn Iversen Mek Verksted, Flekkefjord, Norway, as *Antares* LK491, 1978. Became *Serene* LK491, 1985; *Serene* FR491, 1987. Sold to Icelandic/Mexican consortium, 2006.

Regent Bird BCK110, built by Richard Irvin & Sons Ltd, Peterhead, as *Graceful* PD343, 1961. Became *Regent Bird* PD343, 1974; *Regent Bird* BCK110, 1988. Wrecked at Orkney, 14 January 1995.

Sophie Louise WY168, built by James N. Miller & Sons Ltd, St Monance, 1988. Became *Ocean Harvest III* FR145, 2002; *Sophie Louise* WY168, 2013; *Harvest Moon* FR366, 2016; *Sophie Louise* LH177, 2021.

Lothian Rose LH380, built by McTay Marine Ltd, Bromborough, Cheshire, as *Lothian Rose II* LH380, 1979. Became *Lothian Rose* LH380, 1987. Moved to Eire and became *Lothian Rose* S495, 2000.

Challenge FR77, built by Fiskerstrand Verft, Fiskerstrand, Norway, as *Zephyr* LK394, 1980. Became *Challenge* FR77, 1996; *Alert* FR777, 2002. Moved to Northern Ireland 2006 followed by sale to Sweden. Became *Ganthi VII* GG205.

Rivo I K391, built by Steelship Ltd, Truro, 1986. Decommissioned 2002.

Acorn INS237, built by Johs Kristensen Skibsbyggeri ApS, Hvide Sande, Denmark, 1980. Decommissioned 2022.

Fruitful Bough PD109, built by Macduff Shipyards Ltd, Macduff, 2004. Became *Jenna Amber* PD223 2019; *Gracious* PD103, 2019.

Lunar Bow PD265, built by Simek AS, Flekkefjord, Norway, 1996. Moved to Iceland and became *Asgrimur Halldorsson* SF250, 2000.

Lunar Bow PD265, built by Simek AS, Flekkefjord, Norway, 2000. Moved to Iceland and became *Asgrimur Halldorsson* SF250, 2008.

Adenia LK193, built by Astilleros Zamakona SA, Santurce, Bilbao, Spain, 2019.

Celestial Dawn BF109, built by Macduff Shipyards Ltd, Macduff, 2000. Became *Claire Marie* BF101, 2021.

Unity FR165, built by Sigbjorn Iversen Mek Verksted, Flekkefjord, Norway, as *Pathway* PD165, 1984. Became *Unity B* FR165, 1995; *Unity* FR165, 1995. Moved to Norway and became *Unity* R-65-ES, 2001.

Faithlie FR220, built by Vestvaerftet ApS, Hvide Sande, Denmark, 2017.

ABOUT THE AUTHOR

Peter Drummond has a lifelong interest in the Scottish fishing industry that he can trace back to the summer of 1962, when he was 3 years old and visiting Portpatrick harbour – all the ring net boats landing herring from the Isle of Man fishery left an impression that has never gone away!

Fast forward a few years, to when he was bigger and could afford a decent camera, and a start was made to his photograph collection, which eventually led to him co-authoring four books on fishing vessels. For this, his fifth book, he is trying something different: a compilation of his best pictures, deriving from a long-overdue indexing of the collection during the Covid lockdowns. As Peter says, honesty requires an admission that camera and photographer don't always function simultaneously – but when they do, the results can be quite decent, and he had a lot of fun seeing how much variety could be created for a book.

Accord PD90 outward bound from Peterhead after landing a catch of North Sea herring.